THE POWER OF CONTROL THOUGHT

To Dr Bellacé

Enjoy the book.

Roy F Messier.

THE POWER OF CONTROL THOUGHT

ROY F. MESSIER

The Power of Control Thought is Within Your Grasp

iUniverse, Inc.
Bloomington

The Power of Control Thought

iUniverse books may be ordered through booksellers or by contacting:

iUniverse
1663 Liberty Drive
Bloomington, IN 47403
www.iuniverse.com
1-800-Authors (1-800-288-4677)

Because of the dynamic nature of the Internet, any web addresses or links contained in this book may have changed since publication and may no longer be valid. The views expressed in this work are solely those of the author and do not necessarily reflect the views of the publisher, and the publisher hereby disclaims any responsibility for them.

Any people depicted in stock imagery provided by Thinkstock are models, and such images are being used for illustrative purposes only.

Certain stock imagery © Thinkstock.

ISBN: 978-1-4620-8205-6 (sc)
ISBN: 978-1-4620-8206-3 (hc)
ISBN: 978-1-4620-8207-0 (e)

Printed in the United States of America

iUniverse rev. date: 04/26/2012

ACKNOWLEDGEMENTS

I would like to express my deepest gratitude and special thanks to my friend André Yan for all his help in making this book a reality.

Thanks to my sister Julie Manzi for all she has done for me throughout the years. Her support was deeply appreciated.

I am grateful to my brother Felix Messier, to my sister Emily Prieskorn and to Rita Titus for all their support. To my friends Donald Beck and Geoffrey Webb, thank you for your support and friendship.

CONTENTS

THE POWER OF CONTROL THOUGHT

BY ROY F. MESSIER

Yes, the Power of Thought is within your grasp. In using Control Thought, the critical thing to remember is that it takes time to control all the thoughts you have about everything in your world. The Power of Thought can change your life. Yes, change your life from where you are now to where you want to be. With the Power of Thought you can be healed of almost anything, and you can reach every goal you have dreamed of.

The Power of Thought is most important in today's world. For without it, you would go through life not fully free. The Power of Thought is such an energy that once you have mastered it, you can live a life full of wonder, a life with control – your control, not someone else's thoughts of how you should live. Up until now, man has only been able to master a small percentage of this power. When you master the Power of Thought, your world will be like a dream of wonderful experiences. With each new day, the unfolding of your life will prove that you are in full control. In any situation you can choose to have thoughts that will govern your environment.

When people around you are in disarray, you don't have to be a part of it. With the Power of Thought, you are in control of what you think. In choosing Control Thought, you will find, when confronted with anger or

1

unhappiness or any other happening, you can be the master of what you do in any situation. In this way, only good comes from your thoughts. You can become involved or you can walk away, choosing whether you wish to be a part of this event. The choice is yours.

No one knows exactly what the mind is. All that is known about the mind is what it does. The greatest philosophers know no more than this, except that they tell us more of how it works. I choose to think of the Mind as the inner spirit, that self that lives within. For when you are in harmony with the Universe and connect with the One Mind, then all thoughts are creative, coming from the One Intelligence. You set the power of these thoughts into motion according to your belief. You have a consciousness through which you believe.

Through your awareness of the Mind, which is the creative outlet, each moment in your life is built upon your thoughts. Thoughts become the law of life. People can bring into experience whatever they desire.

Your awareness of this process and your ability to use it can transform the way you look at life – from health, personal relationships, business, financial and family. For all the physical and the emotional circumstances in your life, you can work with inner causation rather than struggling with outer circumstances. You can produce results. The creative power is the thinker behind your thoughts.

First, one must be willing to allow change to take place. Clearing the path for the new and clarifying your thoughts, then changing them, is a wonderful demonstration of your power. **What you think about you will one day experience.**

Each day is a new beginning. It was never before and will never be again. You have to choose to make each new day for living. You can be miserable or you can change your thoughts and live happily.

You need to have concentration to live a happy life. You have to live your own life, embrace your own experience, and recognize your higher power. You are never alone; there is always a presence within that accompanies you on your path. This presence will always be with you.

Life is forever expanding and moving forward. Life is energy, the world is energy, and the universe is energy. All is energy.

You too have the power to use this energy. Situations can change when you change your attitude and believe that Love is the key, the power that will work for you. You can recognize the presence of this power in everything you do, say and think. You can create a new environment and circumstances every day through the use of Control Thought. Take that great leap forward and use Control Thought to live in the Now.

CONTROL THOUGHT

It's learning to listen to what kind of thoughts you are having and controlling them. Are they positive or negative? All thought inevitably tends to create its physical correspondent. Thought can be changed at any time and thereby change your life.

The thoughts you want to think about are as hard to control as the thoughts you don't want to think about – you can think yourself into being happy or unhappy. It's your responsibility. Your mind is compelled to create the conditions in your life that enter your mind through thought.

When your mind accepts an idea as true, it then *becomes* true for you. Every idea fed into the mind through thought is bound to produce an effect exactly like its cause. You have the power within to make anything happen in your life. You have the same power to change any situation you have created from the thoughts you fed into your mind.

If you do not believe that your mind and your thoughts are limitless in their ability to create the reality of any idea that you can believe in, then you must believe that your mind and your thoughts are limited in their ability. And so your mind then creates experiences that represent your beliefs of limitation. You have seen this work in your own life.

You must run your subjective and emotional life; if not, the world will run it for you and will impregnate it with all sorts of ideas of limitation. Every person must learn to run his or her own life, for no one will take your place when death comes. There is but one power, that which is within.

One true law, and that is your own Spirit. It is the only immaterial power you have; your intuition and inspiration all come from the direction of your thoughts.

Mentalist is the harmonious action of the three most powerful facilities of the mental organization. The first of these is Thought. The second is Ether Energy and the third is Will.

Thought is the intelligence which is collected by the brain, from mental vibrations in harmony with the mind. Ether Energy is the force generated within the brain by the process of thinking. It is through this force that thought travels from the cells of the brain to their respective stations. Thought conveyed by Ether Energy and guided and controlled by your Will becomes a power of such magnitude that neither material nor distance is a barrier to its transmission. The power of man's mind over his body is actually superior to that of any microbe or disease.

You are surrounded by a Universal Mind into which you think. This mind in its original state fills all space. It fills the space that man uses in the Universe. It is in man as well as outside him. As a man thinks, he sets a law in motion, which is creative. Every thought in the life of an individual expresses that individual.

If man wishes to draw greater good into his life, he must learn to master the use of Control Thought. For as a man thinks, so is he.

You can conceive of a greater good than you have so far experienced, so you have the ability to transcend previous experiences and rise above them. Every thought produces a slight molecular change in the substance of the brain, and the repetition of the same sort of thought causes a repetition of the same molecular action, until at last a veritable channel is formed in the brain substance which can only be eradicated by a reverse process of thought.

Self-improvement teachers suggest that if you wish to break an unwanted habit, you should maintain a replacement habit for three consecutive weeks while refraining from the old habit. Many people have found this to be true as they learned to comply with the new habit.

Negative thought patterns tend to repeat themselves with monotonous regularity. But what if you were to replace negative thought patterns with declarations of Divine Truth?

As a result, you would tend to grow healthier, wealthier, happier, more loving and more successful. It can be done by just starting one day at a time and staying with it. Since the new habit is learned, it becomes as automatic as releasing the hand brake of a car. This is so because a new thought pattern or memory groove has been inscribed into the brain cells. It is a universal law that if you reverse the action of a cause, you simultaneously reverse the effect. **Negative thoughts produce negative results. Positive thoughts produce positive results.**

The greatest gift to man is the Power of Thought, through which man can incorporate into his consciousness the mind of the Universe. **Things may happen around you and things may happen to you, but the only things that really count are the things that happen *within* you.** You can't always control what happens to you, but you can control what you think about what happens and what you are thinking at any particular moment.

Suppose you have a sudden fearful thought – about catching a cold, perhaps, or about the security of your job – deal with this thought and dispose of it immediately. Don't procrastinate and don't give in to the thought.

All contagion begins in the mind. There is a moment when you can say "NO" to every physical affliction that comes to you. Deal with the fear quickly, affirm "I am a spiritual being and therefore I will not accept this thought of weakness."

All lack, financial difficulty and illnesses begin in the mind. The danger is when you let these fearful and worrisome thoughts simmer in your mind. You must take action immediately, stand up and speak the word of peace to the storms of the human thoughts. You must remember that all you can ever do is to support or suppress each other. You can only engage in unity or division.

Inevitably whatever you promote, you reproduce. You should aim to engage only in unity. Growing is actually a letting go process of all the thoughts, beliefs and fears that you use to hold yourself back from experiencing the magnificence that you are and that life is.

Your subjective mind accepts your conscious thoughts and works on them, giving them form. Your subconscious mind can store ideas within it and can release them at some later time. This is why Control Thought is so important. You must not accept thoughts you do not want to express. You must reject unwanted thoughts the moment they appear. Block out that which is undesirable. Affirm there is nothing in me to attract negativity. Cause and effect are inseparable, and you can change the effect by changing the cause.

For example, if you have a health condition, you could change the effect by changing the cause. You can heal the condition by coming to terms with the self. Success is a state of mind. Treat the attitude, not the condition.

This is why you must cultivate a positive affirmative attitude of well-being and create confidence in yourself. You are one with the source. Learn to be specific. This gives your creative power direction so your thinking will come to pass. First you need to think what you want to happen. The law of cause and effect is very simple and very sure in its working. You live in a world governed by order.

When you decide to launch into a program of self-improvement, self-realization and the discipline of Control Thought, you will likely experience life with the highest consciousness of love for all. Your knowledge of the truth must at all times lead to self-knowledge; then your own body of knowledge suddenly reveals a new dimension. You must know your own thoughts and train yourself to think what you wish to think, be what you wish to be, feel what you wish to feel, and place no limitations. Think right thoughts until your thoughts become perfectly clear. This process answers every question, solves all problems and is the solution to every difficulty.

As humans we live in a life of dreams which takes place on three planes, the physical, mental and spiritual. The energy of the mind, like other natural energy, already exists.

You should carefully consider whether you are willing to experience the result of your thoughts. If you keep your thoughts fixed upon certain ideas, what you desire in your life will begin to take form. The thoughts that you concentrate on are attained.

We come into this life subjected to group consciousness and to our environment as our personalities unfold.

Everything is thought; nothing moves but thought in the mind, and the only instrument of the mind is thought. When you realize that conscious thoughts operate through an infinite power, then your thoughts can bring about any condition, as perfect as you can conceive. Control Thought, which is built upon realization, has the power to neutralize negative thoughts, just as light has the power to overcome darkness.

We live in a spiritual world which is a perfect universe. So we maintain that we live among potentially perfect people. Your Control Thoughts will understand this and will refuse to believe in its opposite. Let us seek the good and the truth in people and believe in them, even though some people we meet are filled with suffering and limitation. You are dealing with a universal principle through your thought power. Why would you set limits to this? Since Universal principle is the law of the universe, the law knows nothing about limitation of any kind.

At first, children are happy, free, and spontaneous, but as they grow aware of the facts of life, their emotions become more complex. As they hear people talk about death, trouble, divorce, love, marriage and everything else, they begin to react to these emotions subjectively. After a certain age, children have to be re-educated just like adults so that their subjective mind may not reproduce false impressions.

Everything is from within out. Even if you have heard this many times, it's good to keep reminding yourself that whenever you move your mind, everything else in your life will start to move in the same direction.

Create for yourself a new experience by moving away from belief in illness, sadness or limitation of any kind. You will find that when you start to dwell on Control Thought instead of focusing on your problems, these problems will be eliminated from your life. Your greatest gift of the mind is to identify only that which you desire to experience.

However, sometimes you may think, "I'll try," but if you say merely that you will try, you have already accepted defeat. What you are saying is either "I don't know if I can" or "I don't know if I want to." Both of these statements denote doubt, and **doubt is a giant obstacle to the manifestation of good.** Giving up is the opposite of persistently sticking to whatever you strive to accomplish. Deciding to move in a different direction can be like turning the dial on your radio.

With the power of Control Thought, you are in charge of your life and therefore have the choice of deciding which thoughts you will bring to any given situation. Your Control Thought controls the direction of your life. If you think that Control Thought is thinking correctly, then you become a living embodiment of your thoughts. If your thoughts are limited, then that limit is what is available to you. To become a living embodiment of success, your belief in yourself needs at all times to be so firm that it is not short-circuited by doubt, fear or self-distrust.

You can build belief even though limits assume shape and form in the outer world. So re-direct your thinking until you gain new positive beliefs about who you really are. Let go of those negative thoughts you have of yourself and bring into your thoughts only the positive ones. Maybe you feel that you are not living the full, rich, joyous life you intuitively know as your divine right. Perhaps you are unhappy for reasons you don't even understand and want to take steps to achieve greater happiness, but you don't know how to go about it. One way to begin reshaping your life is to adopt new and more positive thoughts and beliefs. When you do this, these thoughts and beliefs are acted upon by the Law of the Mind and the nature of your outer experience is thereby changed.

Thus you have control over your thoughts, with forgiveness of your past, and you see yourself as a perfect and whole spiritual being. A new life emerges for you.

Know that your Control Thoughts establish you toward greater faith and good. Right thoughts govern your affairs, for you have complete dominion over your thoughts and whatever goes on in your life. You have to let go of thoughts of doubt, distrust, worry, condemnation and fear, replacing these negative beliefs with positive thoughts. Know that change for the better is possible for you through understanding and application of Control Thoughts as well as through a conscious awareness of your spiritual perfection.

Remember that happiness lies in the consciousness you have of it. You know that your world is a reflection of yourself. Let yourself make a decisive shift in your thinking. Let yourself move onto a higher level of consciousness, relaxing in the awareness that within you is the total, complete and joyous fulfillment of life itself.

Every time you think, you are thinking into a receptive substance which receives the impression of your thoughts. When you stop to realize how subtle thoughts are, how unconsciously you think negation, how easy it is to get "down and out" mentally, you see that everyone is perpetuating his own condition. This is why people go from bad to worse or from success to greater success.

Only when you gradually, definitely and intelligently take negative thoughts and build them into the structure of positive thoughts can you arrive at the desired reaction. This is called trained thoughts. There is a great difference between conscious and unconscious thought. There is nothing more real than thought. No one has ever seen a thought, yet we see its power to change a person's life. When a thought knows, understands and embodies reality, it becomes reality – and reality is power. When the illusion of opposites disappears from your mental perception, you realize that you are whole. The mind and body cannot be separated. The mind-body interaction is a remarkable force. This power of energy is drawn from everything in life. I've come to believe that whether the power comes from a source beyond me or whether it resides completely within me, I can make use of this power.

People have the power within to dramatically alter the way they think and act. Your experiences are the direct result of your own state of mind and nothing else. You, yourself, are the final arbitrator of your own faith. You are always in the stream of unfolding, and your consciousness is a part of the unfolding of the Universe. Life is here for you. **You are the decision maker, the doer, the receiver of whatever you set into motion.**

Make it a point to remember that there is a direct connection between what is taking place in your experience and your inner thoughts and beliefs. Are you up to date in doing what you have set out to do? If not, perhaps you are feeling a sense of frustration, which not only drains energy but also tends to lower self-esteem. Maintaining a higher self-esteem is vitally important, for it helps you to accomplish your goals successfully.

In what direction is that miracle-working mind of yours moving? Is it fully and firmly sustained in the direction of your choice, or is it perhaps going hither and yon, unformed and unfocused? To experience the good, you also need to develop a habitually positive and joyous way of life in order to combat the effect of negative emotions that may crop up from time to time.

Be willing to completely let go of lesser thoughts in order to experience the greater. Regardless of your present state of mind, recognize that this letting go is not only possible for you but much easier than you might imagine. When you let go of negative thinking and lay hold of reality, then you become aware of your connection with life. You will begin to experience yourself as a success and to expect success. Then you can be the star in your own life!

Control Thoughts can and do create the way you live. Control Thoughts can do anything that the mind can receive from the one Mind, that Infinite Intelligence.

Thoughts create everything that is experienced from the very beginning of your life. Your thoughts are the creative avenue of the mind within you. And since thought is creative, it can heal disease, change the environment, attract friends and cause you to demonstrate success.

Thoughts of happiness create harmony, thoughts of wholeness bring health, and thoughts of abundance make one prosperous. Thoughts produce energy. Energy creates force which in turn produces action and motion. Control Thought through motion displaces the atoms in the air around the body, causing vibrations of thought waves in the atmosphere. After tossing a stone in the water, you will always see ripples in the water. Similarly, everything in the Universe gives off a vibration.

Very few people think original thoughts. That is, thoughts that they draw from the great source of intelligence. As a rule, they think thoughts received from the thought waves of others. Thought atmosphere is what I call thought vibration. Thought is the intelligence collected by the brain for the use of the mind from passing thought vibrations in harmony with the mind. It can enable one to read another person's thoughts. Or by reading the person's action, one can sometimes tell what their thoughts are at that very moment.

MENTAL ATMOSPHERES

Each person has a mental atmosphere which is the result of all he has thought, said and done. Mental atmosphere is very real; its influence constitutes the power of personal attraction. It is almost entirely subjective. You meet someone and turn away, while you are drawn to others without any apparent reason. This is the result of mental atmosphere or thought vibration. The inner thought of the unspoken word often carries more weight than the spoken. Thought-transference, or telepathy, is a commonly known fact. Mental telepathy would not be possible unless there were a medium through which it could operate. This medium is Universal Mind through which all thought-transference or mental telepathy takes place. Telepathy is the act of reading subjective thoughts or receiving conscious thoughts from another without audible words being spoken. Telepathy is the transmission of thoughts. In other words, one might pick up thoughts just as one picks up radio messages if the receiver is in tune with the sender.

If you have felt limited in any way by circumstances in your life, change your thoughts. Use the power of your imagination to see yourself breaking every band and overcoming every obstacle. Know that as one door closes, another door opens, and so you are eternally moving forward into greater experience. As you move into greater awareness of yourself, remember **no one else can do what you have been created to do.**

Each one of us has a special contribution to make to the world through that uniqueness that we are. We have our own special personality. There is no one like you in the whole world. This is the greatest; rejoice that there is no one just like you!

In the Universe of your mind, you are at the very center of the whole Universe, with the divine presence and power. The medium of all thought is the Universal Mind acting as law, and the law is always impersonal, neutral, receptive and reactive.

You should never allow yourself to think of or talk about limitation or poverty. **Life is a mirror and will reflect what you think.** To view limitation is to impress it upon the mind. It is not always easy to run from fear, from poverty and pain, and from the hurt of human existence. But whoever can train himself to do that will be a healed human being. For all healing is accomplished through the power of Control Thought.

Limitation and poverty are not things, but the results of restricted ways of thinking. You are surrounded by a subjective intelligence which received the impression of your thoughts and acted upon them.

Cause and effect are two sides of the same thing, one being an image in the mind and the other an objective condition. Therefore, you must definitely speak your conviction in concrete form. Demonstration means to prove, to exemplify, to manifest, to bring forth, and to project into your experience something that is better than you had yesterday. All words are formed by your thoughts. Energy unconnected does nothing. It is only when it is used and properly directed that it accomplishes things.

Your Control Thought is energy that must be properly directed in your everyday experiences. Every idea through thought in the mind is bound to produce an effect exactly like its cause. The responsibility of setting the law in motion is yours, but the responsibility of making it work is inherent in its own nature.

It is good to become an observer of your thoughts, to see which thoughts are based on pre-judgment, fear, prejudice and lack of trust, and which thoughts are open, loving and peaceful. If you dig deeply and honestly enough, you will find that all thoughts are from one of these sources.

Through your own consciousness, you empower your thoughts. You give birth to your external world. Nothing is done to you except that which you allow or create. Therefore, focus always on the internal world. Most people are too serious about the external world.

Every thought system centers on what you believe you are. If the center of the thought system is true, only truth extends from it. Change in motivation is a major change because the mind is fundamental.

Thought is an invisible substance that materializes in the space of the brain. The cells of the brain take the thought and convert it into energy for the mind. In the mind these thoughts are energized with electrical current. The thought wave is then sent out in the tone and vibration as strong or as weak as the power behind the thought.

Thoughts are words; hate is a thought and love is a thought. There are people who are quick to condemn races, religions, customs and beliefs based solely upon their response to a symbol rather than their own experience. This is because as a child you build up a belief system with the thoughts of others. You need to define your thoughts as to what words you will speak.

Labels cause pain, and to judge another is really hiding from yourself. So it is essential to examine each thought so that it will be a Control Thought, one which will bring peace to you. **Words you use, in a very real sense, determine your belief system.** It is the essential human connection. Each one to us is life personalized. We are each living life as a person, therefore each one of us contains all the intelligence and power we need.

Each of us has within an instrument for the expression of life. Each person has the ability and the power to express life in peace, happiness, abundance and satisfaction. The instrument which can be used to bring into life whatever you desire is the Mind.

When you know how to control your thoughts, you can map your own destiny without limit. **The mind is the creative principle of life.** You must watch your thoughts and your attitudes and you will see for yourself how the Power of Life responds to your mental states. You have the choice to judge righteously, focusing on love and goodness. Let your choice be the latter. Almost everyday you hear of disaster, illness, injustice and poverty, and at these times you are called more than ever to see the good in all of life and to remember that God expresses through every individual.

You are constantly thinking, and as a result of your thinking, you are directing your life experience through the activity of the Law of Mind. So to live a happy, peaceful life you need to pay close attention to your thoughts, since the Law is neutral and will respond to whatever impression it receives, whether positive or negative. You must learn to discipline your thoughts, to reject the negative thoughts and turn your thinking toward the positive to truth, peace, health and all conditions that reflect your life. Thought is the beginning and the end of life. You have been given the option of choosing your thoughts.

The infinite intelligence of the Universe will respond to the beliefs that you plant in the creative soil of the subjective mind. You have only to choose and the Law of Mind honors your selection. **Therefore, in Control Thought, make wise and loving choices in consciousness; then the unfolding of corresponding conditions will manifest.**

In order to move beyond any human condition, you need the ability to align all thoughts with Control Thought. A trained thought employs wonderful power for good that transcends all false belief. **With Control Thought you will see order replace disorder, unity replace separation.**

FAITH

The future is a result of what you make of your today. Faith can be viewed from two perspectives. One is blind faith that often helps us through difficult times. Faith is the acceptance and trust in a Power greater than we. The other kind of faith is one of understanding. It's based on the knowledge that the Power we speak of can be used and directed by us as individuals. Understanding faith, we know that it works. Having this kind of faith, we accept it as our companion and guardian every moment of the day.

Faith is a way of thinking, an attitude of the mind and the inner certitude, knowing that the thoughts we fully accept in our conscious mind will be embodied in our subconscious mind and made manifest. Faith is accepting as true what our reason and senses deny.

The Bible says, what thing so ever ye desire, when ye pray believe that ye receive them, and ye shall have them. Mark 11:24

According to your faith, is it unto you.

It is amazing that within you are the infinite never-ending qualities of intelligence, wisdom, peace, beauty, joy and love – everything you would like to experience in life.

Within you are the facilities by which these qualities can be expressed. You have the ability to choose and decide. Through your power to choose and decide you can build the personality and the character you choose to build. These are all mental facilities and spiritual qualities to use, express and experience as you grow in understanding yourself.

Everything you do is in response to a desire. Your faith directs you in whatever you do. Faith is a condition of the mind. And with desire and faith there is a facility called imagination. Imagination is your plan-making department and it is under your direction. **There is no limit to your ability to use imagination. It's an infinite faculty.**

You also have a faculty called inner guidance. If you were to recognize and fully trust it, and if you were to be true to it and use it, there is no limit to where this faculty of guidance would lead you. It too is an infinite faculty. Any picture firmly held in your imagination is bound to come forth. This is the great unchanging Universal Law, which when you cooperate with its intelligence, makes you absolute master of your condition and environment. Through your thinking you accumulate a mass of ideas, and with your imagination you make them into definite form.

You cannot demonstrate beyond your ability to provide a mental equivalent of your desire. You enter the absolute in such degree as you draw from the relative.

Within you is the unborn possibility of limitless experience. You have the privilege of giving birth to it. Affirm the divine oneness within you, act as though you are and you will be. Daily you must control all of your thoughts that deny the real. Formulate an over-all picture of the way you want your life to be, and make this your major premise. **Infinite intelligence leads and guides you in all your ways.**

Perfect health is for you as the law of harmony operates in your mind and body. Beauty, love, peace and abundance are yours. The principle of right action and divine order govern your entire life. Know the eternal truths of life and know, feel and believe that your subconscious mind responds according to the nature of your conscious mind and its thoughts.

All the belief systems that were taught to you in your early days were someone else's thoughts of how they perceived the world. As you learn to grow in the spiritual world, your belief system gives way to a new belief system. The first thing you need to do is remind yourself that you have within you the ability to create for yourself a new set of circumstances. Also remind yourself that you are looking at your world through a window, and you must roll up the shades and let the light in.

The darkness is your muddled thinking and the light is that which shines forth through you.

So if you feel discouraged or frustrated, roll up the shades of negative thinking and let the light shine through you. See the blessings in your life and realize you are one with all of life. Spirit within you is alive, conscious, aware and active. Turn away from fear and frustration. You are capable and worthy of experiencing greater possibilities in your life. Turn from sadness to happiness. **When you draw upon thoughts of love, the more loving you will be to yourself and others.**

The more you draw upon thoughts of intelligence, the more intelligent you become. Keep thoughts of wholeness, and you will express this wholeness as perfect health. To be healthy is to be whole. Thoughts are ultimate energy. You draw confidently upon them, and then you will have the energy to do all that you choose to do. It is a thought of substance, the substance out of which all things are made.

Many people are besieged by unruly emotions, feelings that seem to overwhelm them and cruelly tear them apart. Often they will say, "I can't help the way I feel," but is this true? Are you to be lifted to the skies with flights of good feeling, only to be plunged low by down spells? Are you not able to use your intellect to help you recognize your powerful emotions, think them through and change them if necessary? If you do not desire the image given, you can call forth other, more positive images.

Your mental attitude can be either positive or negative. You can believe in the basic goodness of life, that life is for you, or you can believe that nothing ever turns out right. I suspect that you would like to have a more positive attitude toward life and would like to have a greater faith. Such faith is possible to cultivate.

Since faith is an attitude, it is inwardly created. You choose to feel empowered from within and then you must act upon your newly-built attitude. Turn completely away from conditions and look only at what you want to see. How can you create an image of wholeness and perfection if you spend your entire time immersed in thoughts of limitations and frustration? The realization of your perfection is the most inspired goal you could ever have. You already know how creative you are—your thoughts are always at work fashioning the events that shape your days. Through your negative thoughts you continue to create failure, sickness and low self-esteem. What is your basic attitude toward life? This is an important question to answer since your attitude affects your life so profoundly. Happy people tend to have a healthy attitude of self-acceptance; they feel that they deserve to be peaceful and happy, while unhappy people tend to display the attitude that they don't deserve a good life.

An attitude that is destructive to happiness is the belief that you do not compare favorably with others. This feeling of inadequacy deprives you of the freedom to move joyously forward into life. However, there is a wonderful way to offset this attitude and that is by radiating love and goodwill to everyone around you. A good way to begin is to affirm, "Love flows through me to everyone I meet today." If you practice this, your attitude toward life and other people will transform you into a lighter dimension of thinking and being. You need to be constantly aware of your thought patterns, and if negative thoughts come to your attention, you must resolve them before they have a chance to grow.

When you do not experience the good things you wish for, perhaps you need to take a look at what is inhibiting the flow of life through you. Some of the blocks may be not forgiving a negative situation or people you feel have done you wrong. Habitually making such remarks as "I can't stand" "Isn't that awful" "I hate" – these blocks represent an inflexible attitude and attract conflict to you. If any of these blocks are familiar to you, clear out the cobwebs and sweep up the debris in your thoughts. It can be done and the rewards are great.

Negative ideas and fearful emotions cannot belong to you, nor can they take part in your reality, because they are life-demeaning. Stop being influenced by anything that would distract you from the awareness of the right and perfect power of all your good ideas and great desires and the inevitability of their timely unfolding in your life.

Your consciousness is established in right ideas. It is entirely supportive emotionally, which operates as the law of your life. Here and now recognize that everything about you that really matters is indestructible and inseparable from you. Understand that you are fully alive and fully aware. What you really are is mind thinking clearly.

The infinite spirit within you comes forth into your thinking, feeling and action. It can take complete possession of you and direct you in all your ways and living. **Invite your spirit to be part of your life.** When permitted, this spirit can move through your thinking, your feeling, your body and all your personal affairs. Spiritual means thinking in a way that transcends limited concepts of life. Therefore you can become more spiritual by realizing the truth of yourself, beliefs that will manifest in your life as health, harmony, happiness, abundance, beauty and love.

Spiritual and human thoughts are two ways of thinking: not two minds, but two ways of using the same mind. It is impossible for spirit ever to hand over its creation to some power that is not self.

All forms and circumstances that surround you are manifestations of the creative power of thought. The person who lives spiritually does not live much differently from others, except with fewer problems. There is a way to begin now to become more aware of the spirit in you and in others. The way is open to everyone and requires no special talent. But you must be determined and willing to spend time at it.

You have the power over everything in your life today. The world with its confusion of external events, its collective beliefs and its frenzied movements, need not have an effect on you. Your positive thoughts and ideas direct you into positive action and deeds. Your own conviction gives life to your thoughts as you become the power within.

Each solidly directed thought has all the strength of the universe within you. **Everything you do is part of the infinite process of creation.** Your every thought is made manifest by the power working within you. This power removes and eliminates from your life all ideas of ignorance, prejudice, limitation, age, disease, fear and hate.

Be forever grateful for your awareness of the truth that the infinite wisdom of the spirit within makes clear. Whatever you need to know is made known. So choose to focus your complete attention to Control Thought no matter what, stand firmly on principle and in doing so, you'll experience inner peace.

There is no war in your world because there is no war in your mind. Each day your consciousness of power increases and you enjoy complete dominion. Your life is a celebration of joy because you know that Control Thoughts and right action always will prevail. It's because of man's nature and not his will that his thoughts are creative. This is important, for what you think is what you are instructing the mind to create for you, through you.

Faith creates your thought images and turns them into expressions of how you play the game of life. It's important to remember you are a part of the whole, yet you are not the entire whole. Your reality is part of everything. We are united in the universe but yet separate as individuals.

Everything in the Universe is in accord with the law, just as faith is a law and acts as such. The law of cause and effect transcends a higher use of the law. When faith is consciously applied for a definite purpose, mental and spiritual forces come together as one. Faith applied to specific purposes brings definite results. Deep conviction results when you are clear of confusion. The formula to faith is the realization that natural law is rooted in the Divine Intelligence. **Faith is the law of nature at work.**

All people who have exercised effective faith have used the law. Faith is the way we live. It can be consciously acquired if you work at it.

The Universe honors our acceptance of it. It's our conscious thought patterns that are continually impinging upon our environment.

Faith is a position of confidences, assurances and rest in the mind. The higher the sense of truth, the greater will be the realization. The highest is the power of thought with the unity of the spirit. The action of thought with spirit produces creation. Know that thought can influence objectives, whether in the physical body or the physical environment.

The mind and spirit must turn completely from conditions as they are, and must contemplate them as they ought to be, never as they appear to be. The spirit is timeless. Therefore the mind and spirit must not deal with time. You must transcend time.

Faith of God is very different from a faith in God. The faith of God is God. In your spiritual evaluations this transition will gradually take place where you shall cease having faith in and shall have faith of God.

Your mental faculties and spiritual qualities are to be expressed as you grow in the understanding of faith. In using the principle of faith, know that you can use faith any time you need it; it's always with you. Each one of us is a world within a world at the very center of the Universe within ourselves.

With the absolute that has its creation within you, express this faith. Through your power you have the ability to choose and decide to build the personality and character you desire. Have faith in the infinite possibilities of the individual. **You can map your own destiny without limit.** The possibilities are endless – only you can decide how you want your life to be. Remember there is no limit to what you can do, have or be.

There is a difference between faith and belief. You can have faith with little belief, but a strong belief in faith reaches a point that you no longer have to work at it – it's always there in your thoughts. It is necessary to go through the stages of denial, anger, and depression before you get to acceptance. If you are to live a balanced life and really enjoy living, you need to break away from the daily routine and just be with yourself sometime in the day.

All you need to do is communicate your need to the divine mind and let it take care of the details. This is faith. Be open to good, for you cannot harbor negative thoughts and expect good to flow from you. You are part of the one mind, the intelligence of the Universe. **Remember that inner peace begins with a thought within you.**

SPIRITUAL AWARENESS

While the human mind is impermanent, the Divine cannot be uprooted. It can only appear to be covered up. Spiritual mind practice is an uncovering of the divine nature. You should have a procedure, a way to arrive at Spiritual Awareness.

Spiritual awareness contains the definition of spirit. Its nature is the action of spirit within itself. It operates with the truth of your being. Spirit is first cause, the conscious mind and the power that knows itself, a conscious being. To be self-conscious is to be a spiritual entity. We all live and move in our own world controlled by thoughts.

Spirit is intelligence living in a subatomic atom, going through a world of matter as form. At any time Spirit as form has the power to connect with your mind. The physical Universe is controlled and governed by principles of harmony, unity and peace. Everything in the Universe is manifest by the result of its activity. Pure spirit is at the center of every organ, action and function of man's being. You are using a principle that automatically reacts to you by corresponding with your mental attitudes.

All your thoughts, words and affirmations are for the purpose of bringing the consciousness to a higher level of spiritual acceptance. Your spiritual awareness is the secret place of the most high within you. The spiritual mind is the creative factor. Everything is created out of you from all of your thoughts.

Always remember that Spiritual mind practice is an uncovering of the Divine nature in you. On the other side of confusion, there is peace; on the other side of discord there is poise.

The Universe can only give you what you are aware of. Do not think of the spirit as separate from you, but as within as well as around you. Spirit is the interaction within. It is this Spirit that talks to you all the time through the thoughts of knowing, reaches the highest level, and has the greatest power. It transposes negative thoughts into positive thoughts, mental facts into their spiritual correspondence and stays with the spiritual correspondence until the material fact surrenders its discordant image to a pattern more in accord with the divine nature.

The power of your thoughts can produce miracles, and you can change anything in your life as you release your mind's power. When you understand this, a whole new life opens up for you. By intuition you are brought into contact with the inner light when you come to the understanding that all is love and yet all is law. Love rules through law. Love points the way and the law makes the way possible.

The unity of the spirit, the all-producing spiritual substance, cannot be limited to any one portion of space, but must be limitless as space itself. The greatest force is the conscious power of thought. You exist in limitless opportunities, forever seeking expression through your thoughts.

There is a mighty power stored in the cells of the body, awaiting liberation. Concentrating upon the word power helps to liberate this force so that it may do its perfect work in the mind and body. This power is yours to control. This power creates conditions for the ideas that are fed into your mind and your mind, in turn, creates experiences that represent your beliefs. The energy of any belief you have compels you to behave in a manner that is in harmony with your belief. In the most remarkable manner, you are responsible for creating the way you feel, your health and well-being, the nature and quality of all your relationships, success, joy and fulfillment, without exception.

It is unnatural to struggle or be in pain, unhappy or have any limitation of any kind. If you do, it's the result of a lack of information. When you are informed and know the truth, you are free to live your life joyfully and abundantly, which is the natural state of being. Through your thoughts, the power is yours to challenge these forces; then there are no limitations to what thoughts can do for you. With the desire to do something, success is yours.

Directed faith makes every thought crackle with power that you can raise to limitless heights, impelled by the lifting forces of your mighty new self-confidence. It starts with the willingness to have experience as cause in any matter. Responsibility is not a burden, fault, praise, blame, credit, or shame. All these include judgment and evaluations of good or bad, right or wrong, better or worse. They are derived from a premise in which self is considered to be a thing or an object.

Man as an individual can do as he will with himself. Every thought sets desire in motion on the mind, after which the body reacts through action. Whenever the image of thought is set, there the power to create resides. By giving your complete attention to any idea, you automatically embody it. The thought becomes the thing, and more than thing; thoughts cause things to happen. Your subconscious, existing just below the threshold of consciousness, is characterized by sensation.

It is time to look at your future. Be good to yourself. To be spiritually aware is to put your thoughts in order. It's not enough to know about health; you must experience it. A healthy future is within your reach. That is why it's so important to keep your energy on health all positive.

Self-discipline is the key to success. It takes self-discipline to take control of your thoughts, for with discipline you become the master of what you want in life. When you master Control Thought, you are on your way.

You stand at the center of your experience each day. Whatever you can envision can become a reality. Each day you take yourself wherever you go in life. Remember that each experience you ever had will stay with you the rest of your life. See every problem as containing an opportunity and a lesson.

A little story to illustrate **Stretching Beyond your Limit:**

A daisy looked to its right and then to the left, to the older and wiser daisies and asked the other daisies, "Isn't it possible to become something more than an ordinary daisy?" The wiser daisies said the magnificence of daisies has always been beautiful. They asked the smaller daisy to look around. Together they all looked and said there is no need to be a special daisy. All the other daisies thought that this daisy had strange ideas. But our daisy was committed to becoming something more. It wanted to grow beyond the ordinary to the extraordinary. It wanted to be a super daisy. So it thought and meditated. It talked about nothing other than the potential of a living super daisy.

Then one day this daisy began a new opening at the very center of his existence, radiating an almost spiritual kind of light.

The daisy looked to the right and then to the left and realized he had grown more magnificent than any of the other daisies. The lone super daisy continued to reach toward the sun, to radiate its beauty and to share its magnificence with others. The super daisy knew that its courage, commitment and conviction had burst forth with a whole new possibility for daisies, and that eventually all daisies would become super.

Within each one of us, the story of the daisies and the story of the human race plays itself out over and over again. For within each of us is the longing to become something more.

SUBCONSCIOUS

Consciousness is characterized by sensation, emotion, volition and thought. Subconscious is the unknown that can be changed.

Divine ideas inform all human thought, seeking admittance through the doorway of the mind. Most of your psychic or subjective conflicts rise from a consciousness of being isolated or being separated from your good. You must admit that all thought is creative, according to the impulse, emotion or conviction behind the thought. Whatever can be reduced to a mental state can be changed by the opposite mental state.

You have a vast subconscious power at your disposal, but it can act only in accordance with the ideas or molds given it. If the idea of lack is your dominant mold, your mind has no alternative but to produce lack. Conversely, if the idea of abundance is dominant, the mind will act on this and produce those things that represent abundance to you. So focus on abundance with Control Thought for what you want to experience. Let's stop the old habit of hanging onto what you don't want, and build your belief in the reality of abundance. You can affirm "My bills are paid," instead of "I don't have enough money." Such statements help you let go of the problem.

You need to open yourself to the good you desire. This means you must be willing to give up whatever is keeping you from receiving your desire, such as a belief in lack or a feeling that you aren't worthy. Start now to build a consciousness of belief and acceptance. Thoughts are images that you have made and these thought images go to the subconscious, where they are acted upon.

It is because the thoughts you think appear as images that you do not recognize them for what they are. You think you thought them, and so you think you see them. **Image-making takes the place of seeing and replaces vision with illusions.**

Realize you are one with the Infinite Intelligence of your subconscious mind, which knows no obstacle to your thoughts. Know that the work of the Infinite Power of your subconscious cannot be hindered. Infinite Intelligence always finishes successfully whatever it begins. Creative wisdom works through you, bringing all your plans and purposes to completion. Whatever you start, you bring to a successful conclusion. Your aim in life is to give wonderful service, and all those whom you contact are blessed by what you have to offer. **All your work comes to full fruition in divine order.**

How much do you want what you want? When you know what you want, you will definitely refuse to let the thieves of negativity – hatred, anger, hostility and ill will – rob you of peace and harmony.

You are a goal-oriented being, always seeking to bring into your life the good you deeply desire. Remember that subconsciously you either attract this good to yourself through what you accept and believe, or you keep it from manifesting in your life through your failure to accept it. Your Control Thought of what you deeply desire will bring about the answer to attaining your goals. To aid in achieving your desire, you must fire the imagination with an idea and believe totally in your ability to achieve it. **With Control Thought you can turn on the power of enthusiasm** for whatever your desire may be, and know that this is true. Believe in yourself; believe in your ability to give yourself successfully to whatever is really important to you, and know that this is true. If you are not experiencing the good you desire, you need to change your thinking. Let's say you do not like what you are hearing in your mind or you are not getting what you are looking for; then it's time to change all thoughts that keep you from getting what you want in life.

Whenever you make a demand upon the Universe through Control Thought, out of that very demand is created its fulfillment. But that can only be when the demand is in the nature of the Universe. When intelligence makes a demand upon itself, it answers its own demand out of its own nature and cannot help doing so. There is a Divine Intelligence that knows the right answer and accepts this statement as true. The answer to that problem is right then and there created in the mind.

The starting point to all achievements is definiteness of purpose. Fixing your goals is essential before you can expect results. **What the mind can conceive and believe the mind can achieve.**

Visualize your intended destination, and your subconscious mind is affected by this self-suggestion. **Goal-setting is one of the most exciting and rewarding habits you can acquire.** The process of systematically setting goals allows you to get exactly what you want.

Remember that whenever your subconscious mind accepts an idea, it immediately begins to execute it. This law holds true for all ideas, good or bad. If you use it negatively, it brings trouble, failure and confusion. When you use it constructively, it brings freedom and peace of mind. To know the power of your subconscious mind is the doorway to success and happiness.

The field of intelligence in which you are now functioning is encouraging you to open up, to become aware of different aspects, and to embrace a larger potential. The entire human potential is challenging you and inviting you to embrace the extraordinary being that lies in each of us. Perhaps we are not so much human beings as we are humans becoming intelligent beings, living in a harmonious world. When stretching the mind, you open up a whole new set of possibilities. You activate new possibilities when you make the commitment to work together, which opens new potentials within.

We are living in the information age, and information changes the form of things, impacting us in our daily lives. Information is moving so much faster than ever before.

The truth is that nothing big has ever been accomplished by being small. Nothing new has ever been born by staying the same. Our willingness to do whatever it takes to burst forth with new good in our lives is the essential ingredient in transformation. Your mind will work for you automatically if you are willing to adjust yourself accordingly and establish new habits and a new attitude. The possibilities are endless.

You have to have a dream. Create a blueprint and follow through – that's the most important thing. It all starts with a crazy, impossible dream. But the impossible *is* possible. So have a big wild crazy dream and picture the day when all things are possible. Picture the life you would like to have. You stand at the center of your experience each and every hour. Free yourself of fear of what might or might not happen in the future. Change is always happening. This is a great lesson, and when you accept it, your life will be a lot happier.

Your subconscious mind created your body and all its organs. Every thought which is not in harmony with the universe will result in discord and limitation. Your subconscious mind, in accordance with the universal principle, will bring about harmony within you and in your outward life. Your subconscious mind does not reason or question what you feed it. It merely processes and reacts to what you have put into it. All thoughts, negative or positive, go into the subconscious mind.

Logic and reason are of no importance to the subconscious mind. Logic can be the greatest deterrent to a successful life. Logic asks, "Am I worthy of these goals? Am I really entitled to them?" Know that whatever your goals are, you can have them with or without logic.

Expect whatever you desire; you are entitled to all the good that you can conceive. The quality of being consistent in your thinking is the mental model of success.

We don't really plan for the future. We forget that the future is going to come, one way or another. In the 1960s and 70s, the motto was "Live for today." That was then, this is now. Like the weather, changes come and create havoc. Everyone is striving, grappling, or fighting, trying to find inner peace. The future is all the tomorrows, which do come. You choose your future by your everyday choices. No doubt about it, today's world is full of tragedy and heartache. Nevertheless, you have to think about the future with hope in your mind and heart. You would do well to set your heart upon your future ways, doing so in all seriousness.

If you want to play in the future, you have to pay in the present. Each day is the beginning of your future, for each new day lived in the present – the things you do and fail to do – is the payment for tomorrow. The subconscious mind cannot reason like your conscious mind does; the conscious and the subconscious are differentiated, one being objective, the other subjective.

Remember that what is impressed in the subconscious is expressed through the conscious. It is truly a working miracle, the power of the subconscious mind that controls all the body functions. It works around the clock without stopping. What's really interesting is that the average human being knows that there's a relationship between the mind and the body. It's only in traditional healthcare that we have evolved a system that separates the mind and the body. The system of separate mental health and medical treatment has long acted as if the mind and body rarely interacted. Now we're breaking down the barriers between these disciplines. When we create something, we always create it first in thought form. Thought creates an image, which then magnetizes and guides physical energy to flow into form and eventually manifests on the physical plane.

In today's world of fear and uncertainty, you should have a time every day when you can experience the field of silence, bliss, and the reservoirs of energy that lie within.

LISTENING AND DECIDING

Listening is paying attention to sound, hearing something with thoughtful attention and giving consideration. Learn to listen to the voice within.

Deciding involves arriving at a determination after consideration. Deciding is the process of making choices. All of life is full of choices to make, so listen to the voice within and make decisions that will bring peace.

SOME DIFFICULTIES IN LISTENING

Often you want to listen but can not seem to hear. Even though you know that everyone and everything in your environment is trying to communicate with you, you still cannot hear. Believe it or not, it happens because you do not really want to listen. On the surface you may ask for guidance, but you often have already formed the answer to your questions, so what you are really asking for is confirmation. This cannot be given to you if it is not the truth or not in your best interests. Thus, even when you do not seem to hear, you are being assured that you are still in communication.

If you are experiencing difficulty in listening, you should realize that you have hidden, preconceived answers to your "predicaments" which you have not yet uncovered and therefore you are unwilling to let go and listen. Your inner guide will gladly affirm any thoughts you have that are true because these are in concert with your highest mind.

Guilt and fear cause some ideas and beliefs to be forced below the conscious level where they are hidden away from you. When you are willing to seek out these separated thoughts and have them exposed, they can be seen as illusions. They will return to the nothingness which they are.

Anger, fear and discord always attend conflict as you try to defend falsity. Thus, the search for the truth may seem to be a conflict, but once the truth is perceived, it is experienced as true inner peace.

In reality you are always guided and cared for, remaining in constant communication with your source of knowledge, power and harmony. This is the true meaning and expression of life.

Many of your queries are simply some form of the question "Why is this happening to me?" Life is your attempt to integrate what you see happening with what you think should be happening. You are spending time asking God to explain to you the insanity of the world that you yourself created through your attitudes and beliefs. The real question should be "What blessing or lesson would You have me learn?"

Since all things are echoes of God's voice, be open to all sources.

The truth will stay and the rest will pass by. **Willingness is the only condition necessary for listening.** You can tune your awareness to messages from all sources.

We spend most of our life screening out what comes to us in order to separate the good from the bad. You need to let everything come in and to have your inner guide do the judging for you.

The result of this open attitude is remarkable. First, it eliminates your penchant for pre-judgment and then puts judgment where it belongs, with your guide and not with your five senses and past experiences. For instance, we often judge what people are thinking when they have ideas that seem off-base, especially when they hold a belief that differs from our own. You either rush to change their minds, or you dismiss them as ignorant of what is going on. You feel compelled to use almost any method short of violence to silence or change them. Argument and ridicule are two techniques we often use in these situations.

We must admit, however, that these practices seldom change another's mind, nor does either person go away feeling more at peace or joyful from such an encounter. So your aim should be to allow all thoughts to come into your consciousness without any screening, to allow them to pass through your mind without judgment. In doing this, you will be able to identify with the person holding the thought and to love them for who they really are, regardless of whether or not what they are saying is in agreement with your own thinking and belief.

This technique works. Often you will find that you are in much closer agreement with the person than you had originally thought.

You must stay inside the house, so to speak, and allow the one wiser than you to guard the door. Then all ideas that should come in do so, and those that do not belong go away or simply exit through the back door. You should, however, stand guard on your own with your thoughts and encourage your thoughts to be loving, open, trusting and honest. If you desire your thoughts to seek only the highest path, as directed by your inner guide, you do not have to worry about others.

You may not get the answer to your question right away, but it will come and it will be the right answer. You need to be patient – the answer will come at the right time. You should not judge the message or the messenger. You should show that you accept the answer by acting on it. The answer always is a form of forgiveness for you and for others.

As you ask more and more questions to your inner guide, you realize eventually that you have fewer and fewer questions. In the beginning your question will usually be "What should I do about this?" and the answer is usually "Nothing. It is being taken care of." You learn to ask "How?" instead of "What is happening?" Questions that allow you to be at peace are "What does this mean?" and "What is the lesson here?"

If you truly listen, you will discover that all of life is sending the answer to you and is holding up a placard, so to speak, saying this wonderful thing is happening – you are love and you are loved. The true inner voice is always known to those who desire the truth. It is the still small voice that speaks of love and peace. All worthwhile messages are of the spirit; their form is unimportant. Don't worry about the form, or about the person who delivers it. The messenger can come from a Willie Nelson song, "Live One Day at a Time," or from a five-year-old child, a religious leader, spiritual writings, a politician, the daily news, a billboard, your best enemy, or a dream. If you are open you will know when you hear the answer, and the truth will be known even if it is surrounded with all types of distraction. If in doubt, keep listening. You will always know when it comes to you deep inside. You will feel at peace.

When you are confused, hurried, afraid, doubtful, or fearful, just keep listening. All confusion is of your own making, and you merely need to let go. Once you recognize that you can hear clearly, you will find that life goes much more smoothly for you. All blocks to following this inner guidance will melt away. If the flow is temporarily interrupted, simply silence yourself and listen again, and you will be put back on track. **The feeling of flowing in harmony is a wonderful indication that you are listening.** You do not arrange things, but you do experience them. That is the natural outcome of following your inner voice.

PROCEED AS IF AT PEACE

You do not have to ask your guide for permission for everything. This speaks more of fear than of trust. You must have a desire to do well and to be alert. If your peace becomes disturbed, you should stop and ask for guidance and never force things. You should join your will with God's will. This is letting go. There is an old army command, proceed until further orders, which you can use in your daily life. Your inner guide will alert you if something is off track. You can always assume your inner guide is on duty and that things are going fine. If they are not all right and you sense that they are out of harmony, you will know because you proceed with power and confidence, keeping your antennae raised for constant communication. Fear, uneasiness, pain and slight irritation are all signs that it's time to stop and listen.

Even if you accept wrong ideas, they either seem to go nowhere, or they become such a burden that you put them aside in the end. By letting your inner guide do the sorting out, you save time and enjoy the process more. If you sort out the ideas that are useful, or judge some good and some bad immediately, you may miss some and stop looking before you have found the proper solution.

Once, when I was a practitioner, a woman came to see me for help. I asked what I could do for her. She sat there for a while and then began to speak.

"My husband and I are friends with two other couples. We often go out together. One night while we were getting ready for a formal evening, my husband asked if I was seeing someone. I was surprised at his question. 'No, I'm not seeing anyone. Are you?' He said he felt there was something I wasn't telling him, but if not, just forget it and get ready for the party."

"At the party I saw my husband kissing one of the women in our group. They were unaware of my presence. A few days later I found out that my husband has been seeing her for quite a while – and she is pregnant. I don't know if my husband is the father."

She began to cry and said, "I just don't know what to do. I don't want to lose him."

I asked, "Do you love him?"

"Yes, but what do you think I should do?"

"My thoughts are to have forgiveness for him. Spiritual Mind is the divine nature within you, over which you have control. First, I would like you to think of all the negative thoughts you have about your situation, one at a time, and then turn them into positive thoughts. In other words, get yourself out of the picture and see what is going on around you. Soon you'll be able to see what's right for you to do. Your spirit within always knows what to do, and it will show you the way. Forgiveness is an approach to heal any situation.'

I told her that her spirit could restore whatever was out of balance and that she would find the answer in her heart.

"Peace and happiness are yours."

A few months later, she called and told me that her husband was not the father, and he had stopped seeing the other woman.

LISTEN FOR REASSURANCE

The right answer is never forced upon you. It requires your willingness to want it. Then it is given to you or you become aware of it. You should not force ideas on others. Think of it this way: there is an unlimited supply of ideas. After you have considered ideas, the ones that will work start to grow and become apparent.

DAILY DEVOTION

In order to hear properly, you must desire to change your perceptions of the world. You must accept responsibility for your present perceptions, and be willing to see another way.

It is helpful to set aside certain times each day devoted to the purpose of getting in touch with your inner guide. All efforts made in this regard will be rewarded. In reality, there is only one guide and one source, though it can be experienced in many forms.

The messages from your inner guide are always of love and support. If you see another in need, you must ask your guides how to see that person differently, realizing that what you see in others, you may also see in yourself. If you are confused about how to be helpful, your guide will show the way and provide the proper opportunity.

In reality, you are always guided and cared for and in constant communication with your source of knowledge, power and harmony, which is the true meaning and expression of life.

TO WIN ONE HAS TO BEGIN

For any change to occur, you have to know what to change through your thoughts.

If your circumstances are not what you want, then the corresponding thoughts need to change. As one grows in Spiritual Awareness, one grows in the understanding of God within.

Nothing is impossible. Only in your state of mind can you imagine it as being impossible. The results from your acts may vary, but if it is possible to visualize something, that something is possible. With God's help, all things are possible. So it is in all life's experience. The so-called unpleasant brings the pleasant and shows you happiness, and your appreciation of happiness is that much greater because of having gone through unhappiness.

Know this to be true and work toward it. Listen with assurance; you are being directed even when you are unaware of it.

DECISION

The creative intelligence of your subconscious mind knows what is best for you. Its tendency is always lifeward and it reveals to you the right decision, which blesses you and all concerned. Give thanks for the answers which you know will come to you. **Infinite Intelligence within you knows all things, and the right decision is revealed to you in divine order.** You will recognize the answer when it comes; your consciousness will know.

Decision cannot be difficult. This is obvious if you realize that you must already have decided not to be wholly joyous if that is how you feel. Therefore, the first step in the undoing is to recognize that you actively decided wrongly, but can as actively decide otherwise. Be very firm with yourself in this, and keep yourself fully aware that the undoing process is merely to return your thinking to Control Thought, in which all thoughts are positive.

I have heard so often, "I am not at peace." When you hear someone say this, that person has made wrong decisions about peace. To have peace you have to make the *decision* to have Control Thought of peace at all times.

If you allow yourself to feel guilty about any decision, you will reinforce the error rather than allow it to be undone for you. The more you believe in Control Thought, the stronger it becomes. Everything in life is but an idea, which came from thought. Like the thought of Heaven, it's a state of thought in which sin is absent and the harmony of Divine Mind is manifest.

Through your own unfocused thoughts – wishful thinking, discouragement, and disdain – you wind up frustrated and unsatisfied, feeling helpless. Know that infinite power responds to you through your thoughts. **You have to recognize your mind as the powerhouse, the link to all the power of the universe. Focus your mind on positive images, and positive thoughts will produce happier circumstances.**

All power lives in and moves through your present thought, so action is your present thoughts. This is where the power is. Right now make the decision to empty your mind of every unhappy thought, feeling and belief. Relax and contemplate that which is good and accept it for yourself. Know that in doing so, you are allowing the inner power to flow through you harmoniously and peacefully. With new and vigorous life, walk forth into this new day with a happy carefree heart and mind.

Right now be aware of your connection in life, and you will be open to vital new pathways of happy, prosperous living. Infinite intelligence now will supply you with everything you require to experience a wonderful, uplifting life.

Appreciate the power of Control Thought in your desire to appreciate life. For within you is a power that can deal with any situation. All you need to do is to get in touch with your higher self.

VISION AND DECISION

See your source of being. Vision is an important factor in your progress. It's creative action within the great law. Dreaming either looks backward at what has happened or lives in a make-believe world of imaginary satisfactions. Vision, on the other hand, sees your idea as creative action causing things to happen. **In other words, stop dreaming of what you would like to be or do, and see yourself already doing it.** Only that which you see is included in your experience. See the perfect you, see the self in your action. What you see in the mind you can see as your objective form. See yourself expressing creatively. Pause to enjoy what you see yourself doing. See spirit in perfect action. Visualize success and yourself as successful. Extend this vision. Remember it is already established in your consciousness.

Be definite in seeing what you want to see. Make your project complete in your mind. With each delineation of thought, see your world as already existing. Your vision enables you to act creatively and make your dream become real to you. Right decision is the action of Control Thought.

You can decide today to reject or accept what you have. You must learn to rule your own life. The clearer your thoughts are about your real nature, the more you'll be able to experience it. You are an individual and must express that individuality to the fullest extent of your personality. Never forget the power you have is bigger than anything you have ever created; therefore, you can never feel inadequate. You are in total control of your thoughts, total control of your reaction to life, and therefore total control of your experience.

Illuminate your thoughts through choice and decisions, which then direct this power toward ever-increasing good in all areas of your life. This is achieved through self-work, self- awareness and self-expression. This is a dynamic action. It is action of consciousness unfolding. It is this inner energy that moves through you and heightens your emotion to a conscious, growing awareness of truth. You have to believe that the movement of thought is the movement of your mind. One part is thought, the other is action. You have to look upon the process until you arrive at this understanding. Being spiritual in mind and body, the action brought forth by the mind can do wonderful things. Having control of your thoughts and projecting them into specific conditions will bring the reality of life. **You are the thinker in your world.**

HEALTH

Health is the condition of being sound in body, mind and spirit. It's freedom from physical disease or pain. It is the ability of the body to sustain itself physically, mentally, and spiritually in a wholesome way. It's in every thought that implies a positive contribution to health.

There is a state of consciousness which can heal, and that state of consciousness can be obtained if you want it and are willing to work for it. The theory rests entirely upon the supposition that it is impossible to have a true subjective concept of health without there being a positive absolute and equal objective fact. The two will balance; in every action there is always an opposite and equal reaction. Life is a reflection, which nothing can stop. When you do not create the reflection of bad health, you embody the image of perfect health, which in truth will be the reflection. Each of us has to take charge of our own life in ways that will lead us to experience a deep sense of joy, satisfaction and fulfillment. It all starts with the awareness that the mind can and does control all that we are from the very beginning.

The law of cause and effect operates on your beliefs as you actually believe them. You have been endowed with a creative mind, whether or not you know it.

Unconscious thought patterns are the subjective thoughts which come from your environment and from your everyday thinking. An individual subconsciously believes much about himself that contradicts his true spiritual nature. The game of life is simply a form you adopted, which occupies time and space. There is a place in every man's being where he is One with all because he is One with the Universal spirit which is in all.

Everything is governed by law, and all laws must be the operation of intelligence within itself. Thought is the intelligence which is collected by the brain for the use of the mind. Thought energy is the force generated in the brain by the process of thinking. It's this force that gets thoughts to travel from cells of the brain to their destination.

You are tuned into the healing power that your mind is tuned into. If you continue to think there is no cure for your illness, then there is no cure because you are tuned in on a negative channel. The healing vibrations your mind generates are the great power that does the healing. These vibrations I call a healing channel, but it cannot get through if your thoughts are in disharmony. This vibration is the energy that controls the power to its transmission. The will of man is the operator, director and dispatcher which guide the thoughts to their respective stations for healing to take place.

When you are ill, if you have people around you who are sending you love, the faster you will get well. Realize you have the choice to zero in to the spirit and power; you are your own doctor. Doctors can diagnose your illness, but you yourself have to cure it. If there is a need for greater health in your life, now is the time to examine your thoughts. The law of mind responds to your beliefs and brings about circumstances in your life from these beliefs, whether conscious or subconscious.

Miracles are miracles only because we don't understand the processes involved. The dictionary defines the word "miracle" as an effect in the physical world which surpasses all known human natural powers and is therefore ascribed to a supernatural agency.

Know that the power of your thoughts can produce miracles and that you can change anything in your life. As you release your mind power of understanding, you can heal when you affirm light because enlightenment is understanding in expression. By intuition you are brought into contact with the inner light.

The conscious use of understanding as it operates from the mind affects the entire body. The starting point is in the realization of the One designated as the Almighty, that which is a presence dwelling in you, a force surrounding you and a principle by which you live.

With this concept you'll find yourself caught up in a new consciousness that will change your life. You have only to open your mind to accept that which is within you. The law of the Universe for you is a perfect life, wisdom, love and health. You can be healed when you make contact with your inner being and have faith. When you are conscious of a perfect life, the body is whole. You must not focus on the imperfect, but remain conscious of the perfect alone.

Penetrating intelligence forms the condition for cosmic order. An enlightened person, therefore, positions himself with the cosmic focus. Good physical health is first of all a state of mind. If you desire to be healthier than you are right now, then start doing your part to consistently and immediately reverse all negative thoughts of sickness and disease. When you do so, you will be on the pathway to feeling better. Now, this doesn't mean that you should stop seeing your doctor, for the doctor is part of your healing process. What it means is that you are providing another avenue through which healing can more readily occur. Spend time each day meditating on the wonderful working power of thought in action in and through your body – every atom, cell, organ and function – and affirm **EVERY CELL IN MY BODY IS DANCING WITH JOY AND I AM ALIVE WITH NEW AND VIGOROUS LIFE.**

Live your life as you would like. Do not look outside of yourself for the answer to your quest.

The greatest mistake lies in man's looking outside himself for his future, whether it's health, success or love. You determine your own future. You can set your Control Thought in motion to make all the change you need through your own thinking. You can always move from where you are to where you desire to be. Start today to experience the best health which can be yours. Every thought you think will take action in the outer world. Thoughts are the building blocks you are using to create your health. As far as the mental law is concerned, you break the law every time you think in negative or destructive ways. Your thoughts are your life. The sooner you believe this to be true, the better the world around you will be.

You may keep all the commandments in the Bible and still experience the judgment of Hell's fire. When Jesus refers to the judgment of Hell's fire, it is the punishment for your inner conflict that leads to physical stress, pain and disease and to the human problems of success and failure. Thoughts of weakness keep the image in the mind's eye. Disease and limitation are neither person, place, nor thing; they are only the thought power that you give to them. Limited thoughts are opposite to the realization of health, happiness and harmony.

Never limit your view of life by any past experience or belief. If you have confusion in any condition, then there must have been confusion in the thoughts of that condition.

Every pain you have is the result of some fear or lack, some belief in limitation. Know that the power you are using is the power that has always been there for you to use. Let no fear come into your thoughts and if so, take action to show that you no longer have fear of anything. Disease is not to be feared. Disease is real, and your thoughts about health have to be positive all the time. Disease is not always due to conscious thought. While most diseases must first have a subjective cause, this subjective cause, nine times out of ten, is not conscious in the thought of the person who suffers from the condition, but is largely the result of certain combinations of thinking.

A spiritual man knows that health is a reality, and when obstructions occur in any condition in the body, he knows how to take care of the problem. Mental healing is the result of clear thinking and logical reasoning which presents itself to consciousness and is acted upon by the mind. It is a systematic process of reasoning.

Thinking sets caution in motion. Right thoughts constantly pouring into consciousness will eventually purify it. I think it is safe to say that we all want to live healthier and happier lives; however I wonder how much time we actually spend in thinking healthily? Do we really make the effort to improve the quality of our life? In order to live fully, we need not only a healthier body to live in, but also a healthy attitude and a healthy spiritual outlook to complete the circle.

A popular idea seems to be to let someone else do it for us. But no one can make you healthy, nor can anyone else make you happy. Being healthy depends on you. Are you thinking healthy thoughts, or are you thinking and talking about colds, flu, aches and pains? Whatever you focus on in your thoughts increases. You can raise yourself from weakness to strength. You have within you the greatest physician in the world, the power of the mind that causes your heart to beat, your muscles to strengthen and your body to heal. Praise your body, your mind and your emotions. Feel good about yourself. It is because of man's nature and not his will that his thought is creative.

Know that you are filled with an exuberant belief in the Power within you to bring about all the good you can envision, allowing life to flow freely and abundantly through you. Know there is no limit to the full expression of your given potential. It may sound rather harsh to say that what you are experiencing in life is what you believe, but it is the truth. The law of the mind responds to your beliefs and brings about circumstances in your life that support these beliefs, whether conscious or unconscious. For example, if you believe that you are highly susceptible to illness, your body will likely react by manifesting illness. If you believe that chilly wet weather will give you a cold if you are exposed to it, then the law of mind will probably produce a cold for you.

The wonderful thing about the mind is that you can change your input to create a different result. Sounds simple, doesn't it? And it is, but if your beliefs have been held for years, what you need to do is program wholly new assumptions and be faithful about practicing your new patterns of thought. It is possible to revitalize your body. Take time during the day to close your eyes and visualize yourself as strong, healthy and vibrant. Experience what it feels like to be in vigorous good health. Quietly contemplate the truth that there is perfect life within you now. Know this life flows in harmony and in balance throughout your mind and body. Know that you are living in complete wholeness today. Understand that health is a mental as well as a physical state. The condition is a reflection of consciousness; the consciousness is the cause which reflects this condition.

Much emphasis is placed these days on the kind of physical condition we are in. I wonder if we give enough attention to our mental condition. We inhabit two worlds, the outer world of affairs, and the inner world of the mind. Our outer lives are the result of our inner beliefs and convictions. What we think, we experience; what we believe, we establish; what we persistently feed into our mind translates into outer circumstances.

If you are not pleased with your outer world, then you must necessarily take a look at your inner world. If you want something better, something different, now is the time to shape up those mental muscles. **The secret of getting in shape mentally is to clear your negative beliefs and limiting ideas.**

When you do so, you reshape your destiny. This calls for a commitment on your part, but you must believe that the goal of improved circumstances is worth the discipline you need to get there. Stretch your mental muscles and reach into a livingness you have only imagined up to now. Be enthusiastic and zestful in your mental shaping. Release all your negative ideas from your thoughts and replace them with new energizing, positive thoughts. There is a great difference between conscious and unconscious thought, for trained thought is far more powerful than untrained thought.

In training the mind, remember that you are trainable with the use of The Power of Control Thought in every thought you have. In the long run you will be training your mind to think only positive thoughts.

Give thanks for your body, which gives you health and energy. Ponder the miracle of your body, which controls your heart and enables you to live and think of all the separate parts of the body. One of the first things that you must realize is that the Universe is not divided against itself. The good you can experience is equivalent to as much good as you can conceive. When you declare that there is an infinite intelligence governing your life, you begin to believe your thoughts.

When a destructive thought enters and says that you are just fooling yourself, it may be just an old idea declaring that this whole thing is too good to be true. But you are dealing with a power that actually is, and it will be to you what you believe it can be. Doubt will gradually lessen its hold until the time comes when it is as though it never had been. The belief that a thing is too good to be true arises from the subjective atmosphere of your previous experience, which has so limited you that fear maintains a hold on your unconscious imagination.

The best you have ever conceived is but a fragment of your brain power. In fact even the most alert person is never wholly awake, much less fully in action. Why is this? It is because the brain's physiological products are so organized that almost from birth you are continuously blocked by conflicts among internal factions. Yet it is here that you stand on the threshold of a new kind of life through The Power of Control Thought.

The creative potential of the human brain is housed in a potentially indestructible body, which has a built-in replacement system, as long as the supply system is intact. The brain continuously takes itself apart and puts itself back together, not merely cell by cell, but molecule by molecule. Potentially, therefore, the brain is constantly renewed and never ages. It has many functions. Many researchers think of the brain as a sort of a computer, but this comparison is inadequate due to the almost infinite complexity and flexibility of the human brain. We don't just use or depend on our brain; we are our brain, even though it's only two percent of our body weight.

The brain is a voracious consumer of energy, which comes from our blood oxygen. The brain sends millions of flashing signals that carry a load of information we use all the time. We are a perfect entity, living in a perfect Universe.

You have to know that the all-powerful spirit is ever available to healing any discordant condition of your body. You have only to look inside yourself to find your spirit, for it is dwelling in you right now and has always been there from the very beginning.

To the spirit there is no incurable disease. As long as a person is alive, the cells of the body respond to care. You know that disease is largely a state of mind. You know that your thoughts are constantly changing, forever taking on new ways of expression. Through Control Thought you release all your negative thoughts concerning your health. It is time to uproot your thinking about the past and think only positive thoughts to put into your conscious mind. Only Control Thoughts of loving sustain your health. Researchers claim that people who have confidence and a combative attitude suffer fewer recurrences of their disease than people who feel helpless or hopeless. To have confidence in your thoughts, you first must release all of your old beliefs. You cannot be healed from a problem you continually describe. **The perception of wholeness is the consciousness of healing. The body is healed as the inner Mind is transformed.**

The process of regeneration helps explain the healing process on which your own healing and well-being depend. It brings you face to face with the secret of your inner self. Because you are a spiritual being, you can overcome, you can succeed, and you can be healed – because you have the Power of Thought Control Thought

The only action of the Mind is thought, and it is for this reason that your thoughts create external conditions. Life as you know it on this plane is a form of energy which has to be guided and directed in your body by your intelligence. Your body is composed of trillions of atoms, which means that there is intelligence throughout your body. There is a recent belief that there are brain cells throughout our physical being, which means that there is intelligence throughout your body.

Disease is not so much in the body but in the mind which envelops the body. That disease is not so much in the nerves and glands, but in the brain cells located in those nerves and glands. You do not live in your body as much as you live in your thoughts and feelings which envelop your body. Health is present; it does not come and go. It is your awareness of it that comes and goes.

How wonderful are the works of the creative intelligence within you! You can claim that the healing presence in your subconscious is flowing through you in a harmonious, healthy, peaceful way.

Think of that force as a living intelligence, a lovely companion. Firmly believe it is continually flowing through you. Your body and all its organs were created by the Infinite Intelligence within your subconscious mind. It knows how to heal you. Its wisdom fashioned all your organs, tissues, muscles and bones. This infinite healing presence within you can transform every atom of your being, making you whole and perfect when you ask for the healing of any part of the body.

You may ask, "What can be healed?" I would answer, "Anything and everything." If you view healing simply as putting back into balance that which is temporarily out of balance, you understand that the concept of healing is relevant not only to physical disease, but also to all other areas of experience. The way to healing any problem or condition is to bring into balance the underlying pattern of thought and belief which is creating it. I'm telling you that you have to become aware of any negative thoughts and beliefs you may have and eliminate them, replacing them with a complete acceptance of your wholeness. When you thus accept balance and harmony for yourself, undesirable conditions in your life tend to disappear.

Do not worry about the past, be not fearful of the future, and be supremely concentrated in the present, and the right response will come to you in every situation. The field of pure potentiality inside you then becomes all powerful, creating anything it desires.

SELF-REALIZATION

When your identity comes from the self, you keep your energy to yourself, and you experience peace within. To have a zest for life is to appreciate life. To understand life is to know that power is in the present moment. A good life is one with a good beginning and a good ending. You are aware of pleasure and pain, birth and death, health, disease, youth and age, when you identify yourself with the body and all its working parts. As you learn to accept success for yourself, rather than believing that you cannot succeed, believe you will be successful, for whatever you totally accept and believe in becomes a part of your life experience. You have the ability and the power to do whatever it takes to be happy and successful.

Self-realization is gained by simple belief in your own uniqueness as a human being. The truth about you is this: you are not inferior or superior – you are simply you. You are a personality, a unique individual, for there is no one like you in the world. We know that the background of human thought is largely negative, a denial of a harmonious spiritual universe. In healing yourself, you must believe in the power to heal. That healing power is within. You must believe in what you are saying to your body. Mental healing is the result of clear thoughts and logical reasoning presenting itself to the consciousness and acted upon by the mind. The body is healed as the inner Mind is transformed by a renewed image of truth. Your thoughts are the creative avenues of the Mind within you, and since thoughts are creative, they can heal disease, change environments, attract friends, or bring you success.

Thoughts of happiness create harmony; thoughts of abundance make one prosperous; and thoughts of wholeness bring health. When your Mind's power expresses as light, it has a tremendous healing power. Visualizing light is essential to fighting false growths that are considered malignant or cancerous. Picture light in your mind and then bring that light to the part of the body you intend to heal. In doing so, you release a healing power of the light. Keep the light focused on that part of the body for a few mimutes. Do this often – soon you'll see the change.

Thoughts of injustice and self-pity, as well as frightened thoughts of disease and death, can obscure the light within each cell. **A peaceful state of Mind is a healing state of Mind.** You must decree that you are under the influence of Divine understanding and intelligence.

There is no condition that cannot be overcome. Hold to that conviction – believe it, affirm it, know that the more you hold to the spiritual idea of health, the stronger you become in faith. The more you affirm the idea of health, the more you gradually make a channel for the healing power. The more you use The Power of Control Thought, the more aware you will become of your spoken words. You will talk less and think more. We are all connected in the Infinite Intelligence, and this Intelligence has to be tuned into your Mind, which learns and becomes willing to receive its ideas.

You are not alone in experiencing the effects of your thoughts. We all experience our own thoughts. Thinking and its results are really simultaneous, for cause and effect are never separate. The fact that our minds are joined with the Universal Intelligence is initially hard to grasp. At first, since it seems to carry with it an enormous sense of responsibility and may even be regarded as an invasion of privacy. Yet it is a fact that there are no private thoughts.

Behind every effect there's a cause, so it follows that whatever the thought, the physical effect will resemble it. Ideas are real, having the power within them to be made manifest. **Persistent, constructive thought is the greatest power known and the most effective.** If the visible effect in your life is not what it should be, if you are unhappy, sick and poverty stricken, you know the remedy. The truth is always the remedy. When you reverse the process of thought, the effect will be reversed. It's not easy, but you must do it.

As the truth draws upon the subjective state of your thoughts, it stimulates it into newness of action. Everything works from within out. The body is a reflection of your thought. You need to seek the truth and the cause from within. It can be found in no other place. As conscious thought pours truth into the subjective channels of creative energy, you are to inject thoughts of liberty, freedom, health, harmony and success.

RETIREMENT

Although retirement may involve leaving one's occupation or position, actually no one really retires, for your life keeps on going, and each day you need to be involved. Your thoughts about retirement are most important; you must keep the door open to the new. Success is the degree to which you use your faith, attain wealth, or desire to be successful in life. Abraham Lincoln said, "Always bear in mind that your own resolution to succeed is more important than any one thing."

According to Frank Lloyd Wright, "The thing always happens that you really believe in; and the belief in a thing makes it happen."

Retirement is a new venture, a new challenge, a new path, the beginning of a new way of life. Nothing will be the same. Sometimes you may wish to be back at work. That is because you have not thought of what you would like to do with the time on your hands when you retire. That is why it is so important to plan what you will do when that day comes. Looked at this way, retirement offers the joy of learning and a deeper understanding of life. It's another step forward on the ladder of life, and that step is increased wisdom. In the first part of your life's ladder, you are concentrating on education. On the second part of the ladder, you were concentrating on making a living and interested in life all around you – happy and gay, never thinking about the future, just making money and having a good time. But before you knew it, retirement was at your doorstep.

Realize that you are on an endless ocean of life. Use retirement as a time to get interested in the laws of life and the wonder of your subconscious mind. Do something you always wanted to do. Study new subjects and investigate new ideas. When you retire, you have the power to create whatever you would like in your life.

With The Power Of Control Thought, focus on the "I can" and the "I can't" will disappear as you claim your victory. Start now to think you can and believe you can. Move from the impossible to "I'm possible." Become enthusiastic over creative ideas, and continue to learn and grow. In this manner you will remain young at heart and your body will reflect your thinking at all times. Retirement is not the end, but the beginning to live life freely, mentally and physically. Each day you are growing in wisdom and understanding of life and the universe through your studies and interest. Feel the miraculous healing, self-receiving power of your unconscious mind moving through your whole being. Know and feel that you are inspired, uplifted and rejuvenated, revitalized and recharged spiritually. You are a part of the Infinite life, which knows no end. You are a child of eternity.

What you picture inwardly you now express outwardly. Discipline yourself to accept only the best, knowing that as you do, only the best takes place in you. You might like to ask yourself, what attitude do I have toward my life? Do I have the feeling that I can barely cope with my present situation, or am I exhibiting a positive attitude? Your attitude in life can either make or break you.

For every attitude contains in it the seed of a corresponding thought or experience. As an example, consider the attitude of fear. To overcome fear, whatever it may be, say slowly, quietly and positively, 'I am now mastering my fear, I am overcoming it. Now I am relaxed and at ease and I know that it is so." And believe what you are saying. As these positive seeds of thoughts sink into the subconscious, they will grow and you will become poised, serene and calm. The fears that you had will disappear.

When you affirm positively that you have mastered your fears, a definite decision in your conscious mind releases the power of the subconscious. Change fear into faith as these new thoughts come into being, with the use of Control Thought. All things work together for your greater joy, health and prosperity. Know that with Control Thought new ideas are now in your consciousness, and you are completely receptive to them. They sustain you in right thinking and cause you to act with wisdom and love in every situation.

Most problems are immediately solved by using Control Thought, which lead you to the right action. You are receptive to the larger mind, of which you are a part. Its power is your power, its wisdom, its nature, its availability and its willingness make up the creative impulse in your thoughts. Control Thought has the ability to cause wonderful and rewarding experiences in your life, once you decide to create the best in life and the best in yourself.

Use this power to unveil your life as great. Live in the Now. You are continually growing in self-awareness, regardless of what appears to be. You are free to change and to experience your limitless potential. Love and appreciate yourself, that uniqueness that you are. Choose to make the most of yourself and to live your life fully and richly.

Go about the business of making the changes needed to grow and enrich your life. You are what you think. What do you think? How orderly is your thought process? How straight is your thinking? How clear are your thoughts? Answer these questions the best you can, and you will learn what you need to change.

Emotion and reason should be in balance in your life. The science of reasoning or accurate thinking is called logic. Cobwebs in your thinking and wrong premises – expressions such as always, never, anything, everyone, no one, can't, and impossible – all produce a tired body and a very negative attitude.

RIGHT THOUGHT, RIGHT SPEECH, RIGHT ACTION

To have Right Thoughts, you should always be at peace with yourself and the events around you. So often we think from the perspective of prejudice or ignorance. You get an idea that a person is bad, and therefore that all he does must be evil. All men have in them some evil as well as some good.

But unfortunately, it is our custom to fix our attention on the evil and forget about the good. When we do so, we are only looking at one side of this person and ignoring the other side. With Right Thought we might give the same encouragement to the good side of that man's nature.

The next stage is Right Speech. And here again we find the same two divisions. The first is that we should speak always of good things. It is not our business to speak of the evil deeds of others. Even if the story is true, it is still wrong to repeat it, for you can do no good to the man by repeating that he has done wrong. The kindest thing to do is to say nothing about it.

Right Action: three steps necessarily follow one after the other. If you **think** always of good things, you certainly will not **speak** of any evil thoughts, only the good thoughts in your mind. Then the **action** which follows will also be good.

Action must be prompt and yet well considered. We all know some people who, when an emergency arises, seem to become helpless and don't know what to do. Others plunge into action without thinking. Learn to think quickly and act promptly, and yet always with consideration.

SUCCESS

The secret of success and the key to happiness while you are living in this world is to be one hundred percent alive, filled with joy and enthusiasm, a good attitude, and with energy and power. There are many rules of happiness, the most important one being to accept your circumstances. Believe that they are the most appropriate circumstances for you at this particular time. There is a good reason for you to be where you are. You may have something to learn, so rather than feeling unhappy about the situation, look for the lesson in it while you take steps to change your circumstances for the better.

Cultivate the habit of using Control Thought in the way you think about your circumstances. Know that happiness is not the result of acquiring any particular thing, such as wealth or status, but is the result of recognizing your true nature.

The source of true happiness is always within you. Your happiness does not depend on outer conditions or people. See all your experiences in life as moments in the evolving individual expression of I AM THAT I AM.

Control Thoughts constantly poured into consciousness will eventually purify it. Consciously repeat all Control Thought in recognition of that which is true. You need to understand what you are thinking about at all times. For example, if thoughts of negativity come into your Mind, know that there is a Power within you that can change these thoughts at once to positive thoughts.

People have thoughts of something that they want – for example, a new car – but then in the next breath they say "but I can't afford it." Similarly, if they want to lose weight, they think it's not possible for them due to whatever negative reasons. The ultimate result is that nothing happens. Remember that every thought you have can be changed to a more positive thought, and once you start with Control Thought, you need to maintain the process 24 hours a day.

There is a power that acts on all your thoughts. Negative or positive, the power that is in you does not care if the thought is positive of negative. You are what your thoughts are, and all your experiences come from these thoughts. Accept responsibility today and commit yourself, for with every circumstance there is always a door for change. Your reason for being is universal life, which is your origin.

Your life is determined not so much by what happens to you as by the attitude you bring to your life. Circumstances and situations do color your life, but you have been given the mind to choose what the color shall be. Choose to be happy in spite of problems. Choose to practice being happy. Choose to make someone else happy. Continue to improve your ability and realize your vision and values through effective use of Control Thought. Getting what you want requires courage, fortitude and perseverance.

Research suggests that emotions, not IQ, may be the true measure of human intelligence. When we think of brilliance we think of Einstein; high achievers, we imagine, were wired for greatness from birth. But we have to wonder why, over time, natural talent seems to ignite in some people while it dims in others. It is a sign of emotional intelligence. The most visible emotional skills, the ones we recognize most readily, are the people skills that are empathy and graciousness. Like other emotional skills, empathy is an emotional quality which can be seen as a survival skill.

We live in the now, not in the future or the past. What happened yesterday cannot be changed. The only things you can change are forgiveness to yourself and to all others that have caused you pain. So hold on to all the Control Thoughts of living in the now. When you cultivate houghts of love – unconditional love for all – other things in life will fall into place. It is the power of the here and now. To create is a mental operation. It produces ideas, images of perfect whole being. Your mind will find the way to take that idea or image and transform it into realization.

The more spiritual the thought, the higher its manifestation. Spiritual thoughts require an absolute belief in and reliance upon the truth. There is one Infinite Mind from which all things come. This Mind is through, in and around you. It is the only Mind there is, and every time you think, you use it. By the law of mental causation, thoughts always tend to manifest themselves in experience.

To be successful, there is one thing that you need to be clear about, and that is not to judge people, places or things. Everyone has within the power to change hate into forgiveness. **Forgiveness will open up the door to a happy life.** When you hate, you are out of touch with the reality of life that is all around you. Every day we hear of disaster, illness, injustice and poverty, and at times we are called into helping.

Your job is to recognize that something important is going to happen and to see the beauty in the happening. Listen with your whole body and spirit so you can understand, empathize and heal. Your relationships pay off in beautiful moments. You'll have to be extremely diligent about keeping your ego in check now. Allow yourself to be creative in the moment and in time you won't have to second guess yourself. You must be willing to move slowly and carefully to build your relationships.

Human relations sometimes get frustrating when you choose to feel responsible for things over which you have no responsibility. When you see things the way they are, you are able to observe the problems closely and wait for resolution. Focus now on everything that will bring you closer to the person you want to be. There are certain requirements that must be met in order to seize control of any situation. **Awareness is the first step toward recognizing that a persistent negative behavior makes no sense in your life.** You need to change the behavior as fast as you can, which will open up your thinking to a more positive feeling.

Commit yourself to a great amount of energy in any situation. Know that you have an excellent sense of what it will take for the change you want. This is how to make yourself indispensable in your relationships.

Being self-contained can train your ego to fall in line with the dictates of your higher self.

You don't need a lot of acquaintances; all you need is one golden friend. But before you bail on a relationship, be realistic, since you're not going to love every minute of any relationship.

We are all programmed to believe in a certain order of things. Make effortlessness your mantra. You already know what will satisfy your basic needs. You'll get everything you want and more when you stick to your principles. When you learn to pass over your trials and troubles, you learn to control them.

If you're a little baffled as to your troubles, it's probably the result of following some bad advice. Your relationships are like rare gems, and they definitely become even more valuable over time. **Instead of looking for someone special, look for something special in everyone you meet.**

To get the good out of a relationship, you have to delve deeply into it. The effort is well worth it. Relationships have something to teach you, especially the failed ones. Keeping relationships strong requires a super effort.

Figure out your strengths and work with them. If you get stuck, look around and find another way. Don't stop short of a victory. Drop your defenses and be clear about what you have to offer the friendship. See the good in people and give sincere compliments. Doing this sprinkles some fun in your relationships, especially when you are first establishing a friendship. Do it with a light heart.

Love takes time, but that's the easy part – it also takes work, the kind that makes you set your ego aside. Entertain, explore, and fall in love again. Remember feeling is real but fleeting at times. For instance, the flirtatious connection you made with someone interesting might mean something in the moment, but once the moment is gone, so is the meaning.

When you establish a stronger communication line with your heart, you will feel whole in several areas of your life. If your thoughts are of love for family and friends, then you must learn to forgive and forget. In this way you'll be a happy person. This is the highest truth that any human can learn. For without the ability to forgive, you experience pain and disease. When you can forgive and forget, you gain the wisdom of the ages.

ADVERSITY

Use any adversity you encounter as a tool to become sharper, stronger and more loving. Your confidence is partly due to physical improvement and discipline. Look inside and see the kind of person you are and that you show to the world. What kind of thoughts do you put out to the people near you? Ask yourself, "Is this the kind of person I want to show the world?" Are you going through life as if you were only playing at it? Are you the kind of person who wants to control everything? If so, you'll never find happiness. Your relationships with the people you call friends will last only a short time until they get to know you for what you really are.

Life is full of opportunity, but to be successful, you have to work at it. It will never come to you without your involvement in your life's work. **Remember, nothing in life comes to you free; there is always a price to pay for everything you receive.** The future is yours, providing you concentrate on three things you need to be a success – vision, self-confidence, and self-motivation. See yourself as a successful person in your mind's eye. Hold your hard high, knowing you are doing your very best. Self-motivation means you get off the "I'll try" or "I can't" and change the thoughts to *I can do.*

Which needs to be raised more, our standard of living or our standard of thinking? Do you think, or do you just think you think? One of the greatest lessons you can learn is to be your message, to live by your values. While boarding a train, India's great spiritual leader Mahatma Gandhi was asked by a reporter if he had a message to give. Gandhi simply replied, I AM MY MESSAGE.

It was many years ago that I joined Toastmasters and was a member for over ten years. Toastmasters helped me a great deal in my professional life. Toastmaster is a laboratory, a place to practice communication and leadership skills and to learn what works and what doesn't. It is an ideal place to hone skills that reflect your highest values.

To become an interesting person, you must be interested in people, ideas and the world. Read, think and listen to news about other people and their successes. Reading gives you knowledge, not only about your business. Listen to what others say; there is a difference between hearing and really listening.

We are all conducting a personal battle to improve our abilities, and Toastmasters can be of great help. Nothing is as sweet as the thrill of realizing, yes, we have achieved success. **Whatever your goal in life, never give up!** The power of perseverance will always outlast the brightest comet that zooms across the sky.

REACHING YOUR FULL POTENTIAL

There are many forms of success – within your family, your work, friends, love, your children. It is what you yourself see as success, what you need to be happy. As with retirement, you can be happy where you are, enjoying freedom.

The world is full of abundant opportunities. Remember, no one is ever defeated until defeat has been accepted as a reality. A positive state of mind must be the starting point, not mere hopes or wishes. **A burning desire to be and to do is the starting point from which dreams take off. Dream your dreams and then translate them into organized thoughts. Belief in the power of your desire, backed by faith, will bring success through the powerful principle of Control Thought.** You refuse to recognize the word "impossible" and accept nothing as failure. The law of the Mind cannot move unless ideas move in it. Things cannot be projected unless the law projects them. Power lies behind every individual act, through your consciousness.

Tomorrow's success is the result of today's planning. Most failures and miseries can be traced directly to a lack of wise thinking beforehand. Each person's life is individual, and the steps to achievement are not the same for any two people. Your success must necessarily be up to you. **The relationship you have with yourself is the most important one you can nurture.** It will be the basis for all your other relationships. Take the time to get to know yourself better by embarking on a solo excursion.

12 WAYS TO MOTIVATE YOURSELF
TO YOUR FULL POTENTIAL

1. Be specific in the advice you give yourself so you can put it into practice.
2. Establish checkpoints so you can check your progress.
3. Avoid temptation by deliberately avoiding circumstances or thoughts that might sidetrack you.
4. Recognize your limitations. Don't set goals you don't expect to reach.
5. Take advantage of energy peaks, those times of the day when you are habitually in top form.
6. Make an honest distinction between "I can't" and "I don't want to."
7. Get started. Don't stall.
8. Be optimistic and your chances for success will increase.
9. Know how you want to start, what needs to be done first.
10. Use self-signaling devices, notes, cues and reminders.
11. Promise rewards – small rewards for small accomplishments, big rewards for big accomplishments.
12. Give yourself the right to make mistakes. No one is perfect.

The highest objective is the possession of a good character. Character is something each of us must build for ourselves, the total of thousands of things we do daily as we strive to live up to the best that is in us. Dreams are what tomorrows are made of. You can't be greater than your dreams, ideals, hopes and plans.

Success or failure is caused more by your mental attitude than by your mental capacities. The accurate thinker examines the sources of information, weighs statements for motivations, and tests their reasonableness.

Culture is the enlightenment or refinement that results from education or learning from life experiences. It is the enlargement of one's mental horizon; it is not only of individual value, but is also a bond among people of similar tastes.

Self-discipline is the trait you must develop by instruction and self-control. Those who start at a young age and stay with it, develop a habit that will last a lifetime. If you start with Control Thought, you will be on your way to great success.

It all starts with a crazy impossible dream. Remember, the impossible dream is possible. So have yourself a big wild crazy dream and picture it vividly in your mind.

Visualize the life you would like to have; see it in your mind. You stand at the center of your experience every day. Remember to be free of fear about what might or might not happen in the future.

THE KEY TO HAPPINESS IS HAVING DREAMS
THE KEY TO SUCCESS IS MAKING DREAMS COME TRUE

What is security? It's faith and belief. Whatever you think you can do or dream you can do, begin to do it now.

Theodore Roosevelt once said, "Far better it is to dare mighty things – even though checked by failure – than to take rank with these poor souls who neither enjoy much nor suffer much because they live in that gray twilight which knows neither victory or defeat."

HARMONY

Harmony occurs when different accounts interweave into a single thought brought into consciousness or accord. Very few people believe that they draw their thoughts from the great source of intelligence. As a rule, their thoughts are received from the thought waves of others. Whoever possesses strong will power and also has knowledge of the law of mental messages is under a great responsibility to his fellow man, for his thoughts can be sent at will around the world.

The language of tone is the language of the spheres and the language of the universal world. A mental message can be sent to reach only a certain person, or it can be sent to be taken up by a great number of people. The mental organization of every person is tuned naturally to a certain tone or pitch, which may be raised or lowered by choice. While the mental organization of different people usually has different tones, yet there are many people in the world with exactly the same tone, and they all vibrate in unison. So when someone sends forth a message, it is taken up by all who are of the same tone, but it will have no effect upon others. The lowest tone perceived by the human ear is 24 vibrations per second, and the highest is 32,786.

Like electricity, thought waves travel on the ether at a rate of 186,000 miles per second. A person's mental organization depends on the character of his thoughts and the tension or strength of his will. It is always in tune with some people and in harmony with others. In terms of thought, every subject has its own tone or vibration. Love has its own tone, and hate has the opposite tone. One of the secrets of success lies in knowing how to acquire the most beneficial knowledge from others.

A powerful will gives you the ability to throw great energy into a given thought and keep it there until the object is attained. That thought may be for personal action, for the action of another person, or even for a body of people. When your will has gained sufficient strength to make its own decisions, you may then concentrate on attracting thoughts and opinions from the minds of other people, whose ideas may coincide with your own. Select from them the best.

The object in teaching this science to the world is to enable men and women to create mental, spiritual and physical strength. The value of solitude cannot be overestimated; all the great deeds are born in solitude, and all great characters are formed there. All good impulses are stimulated by judicious solitude and concentration of thought. Concentration improves the memory more quickly than any other method. **You should spend time each day thinking in quiet solitude.**

In preparing to send or receive communications, take a comfortable position in an easy chair. Hold your head straight so that the blood supply will not be cut off from the brain. Close your eyes for a few moments to rest them. Following this, focus on some object three or four feet away from you. Then begin to think of the subject on which you desire information or the person you wish to influence. Under these favorable conditions, mental vibrations can be sent out that will reach the brains of others who are in harmony with those thoughts. You can stimulate their thoughts on that subject until they unconsciously send back the information desired. After you have projected your thoughts from fifteen to thirty minutes, you should relax the will and allow the brain to become passive, so that it may receive any new thoughts as they come to you. In this way you will be able to obtain information that will help you to build your success from the minds of others who are in harmony with those thoughts. Have a notebook and pencil handy to write down new ideas. To become a master of mentalist techniques will take some time. Remember that every subject or line of thought has its own tone or vibration. Love has one tone and hate an opposite tone. If you send out thought with vibrations of hate, they will reach those who are capable of a similar passion. The vibration of love sent out will reach people of capable of love.

When your mind accepts an idea as true, it then becomes true for you. In this remarkable manner, you are responsible for creating your life, your feelings, your health and everything else in your life. With each experience comes the lesson, then the growth. **When you change your thoughts, one after another, you can stop playing at life and start living the way you would like.**

All your peace and contentment comes from within. Intelligence belongs to everyone, for our use, but we cannot monopolize it or keep it. Everyone may use as much of it as desired, but we must pass it on for others to use. This realization is the secret of faith; all people are fundamentally the same. **We are all spiritual beings playing out our roles on the stage of life.** The power is given to all from the Creator. The more we use it, the greater the force. We can never deplete this power. This spiritual energy helps us get through many of life's problems that we must experience in order to learn.

LOVE

I t's a strong affection for another arising out of kinship or personal ties. It's an unselfish, loyal affection based on admiration, benevolence, and concern for the good of another. You can promote an attitude of friendship toward everybody and everything. The person who has learned to love all people will find plenty of people who will return that love. Think of the whole world as your friend.

Love is the grandest healing and drawing power on earth. See only the good in people. Refuse to misunderstand or be misunderstood. With your conscious thoughts of love, you will demonstrate your experience. You are dealing with intelligence, and you should recognize the power you are working with. Through the action of the law of Control Thought, love and friendship are attracted automatically to you. The law of your being is the law of the Mind in action, that there is an exact parallel between thought and conditions.

People feel lonesome because they have a sense of separation. The thing to do first is go within to that center of your being and get in touch with the idea of oneness. Begin to feel at one with all of life as it is expressed through all people everywhere.

You can begin by meditating on a statement like I OPEN MYSELF TO ACCEPT THE EXPERIENCE OF FRIENDSHIP. Then gradually you find yourself involved in an atmosphere of lovingness, and people are drawn to you every day.

Another powerful affirmation is I AM NOW FILLED WITH A NEW SURGE OF LOVINGNESS IN MY LIFE. I NO LONGER LOOK OUTSIDE MYSELF FOR FRIENDSHIP. INSTEAD I REACH DEEP WITHIN MY SOUL FOR THE TRUTH THAT I AM A FRIEND TO ALL.

Feel love in everything you do and express love each day, no matter how much love you give to the world around you. You'll never run out, and the more you give, the more you receive. Stop looking for love outside yourself.

Become aware of any negative thoughts that are not in tune with the beauty of nature. You are the conductor of your orchestra, and when you are playing the wrong notes, such as "I don't feel well, things are in a mess, I can't stand my neighbors," or any other negative thoughts you may have, then it is time for you to use Control Thought. If you continue with negative thoughts, is it any wonder that life is not harmonious? You are at the beginning of an eternal journey, and all along the way you can respond to the highest healing power of love. In all forms of healing, let an inner light flow to restore you to your original perfection.

You are at the very center of your consciousness. Love breaks down the bars of thought, shatters the walls of false beliefs and severs the chains of bondage which thought has imposed on you.

Since choosing is related to your thoughts, then thoughts become the effect of your experiences. You can create new thoughts that will be productive so you can act rather than react and choose the feelings you desire.

Spirit has many personalities or facets. Sometimes when you meet someone, you have a strong reaction of either like or dislike. It could be that you had known this spirit before, for the spirit in us can always signal the spirit in another person if they had known each other in another life. I believe that as we go through our many lifetimes, the people that we have known before are always known to us.

Starting today, see everyone through the eyes of love. Use this powerful force to guide your thoughts, your work and your activities. You are at the doorway of your mind and you alone can select the thoughts you have, for you are in full control of your thoughts.

Marriage is a laboratory of individual unfolding, with two people who see the possibilities of working together to bring about a mutual future.

True love is spiritual perception, and ideally, marriage is that as well. Undoubtedly there will also be challenges and tests for you in many areas of your life.

Life is a process of growth as you move from classroom to classroom. You are not the same as you were ten years ago, or even just yesterday. Give thanks that life is lived one day at a time and that every day is a great opportunity for you to be strong, overcome, achieve and to be happy. Remember that love is not an emotion that begins in you and ends in the positive response of another.

Sometimes others can't seem to handle love, or perhaps your will seems to run temporarily dry and you don't feel loving. What then? These are the times to remember that love is a universal essence, ultimately yours to create or withhold. Love is always occurring, whether you are able to express it or not. There is no situation which cannot be molded and shaped by love, however troubled it may appear at the moment.

If you want direction, ask for it. But be specific and willing to receive the answer. Often people say, "I ask but I don't receive an answer." What they could be saying is that they don't really expect to receive an answer. The emptiness you may experience will never be filled until you realize that within you now are the answers to your every need.

Train yourself to become absolutely still and recognize your oneness with your mind.

If an answer does not come to you during your quiet times of meditation, simply get up and go about your business with a feeing that at the right time you will receive the answer. Believe that the answer will come to you. The next step, regardless of where you may be on the pathway of learning, is to release your experience into the world and absorb an expanded sense of life and love.

There are five steps that you must master in your thoughts. They are Love, Faith, Hope, Courage and Wisdom. These are the thoughts that will guide you to your future. This is the art and science of day-to-day living. Give knowledge and thanks for your everyday talents and accomplishments. By using the power of Control Thought, these thoughts will guide you through life in your work and build a good character. **Remember that love is at the center of your being.**

Love is a four-letter word, but it is so powerful a word that it can change a person's life. Each moment is spent in the Now. Now spend that moment in total Control and that moment will be filled with love. Remember love is at the center of your being and keeps open the channels of communication.

Faith and trust in a power greater than you, allegiance to duty or a person, a belief in something for which there is no proof requires complete trust. Faith is a way of thinking, an attitude of the mind. Faith can be your companion and guardian throughout your life.

Hope is a desire and belief in fulfillment.

It takes courage of mind and heart that you shall not be afraid to travel where there is no blazed trail. You shall not hesitate to sacrifice when by sacrificing, you can contribute to others. You must learn to have courage to rule your own life and choose to react to life in a healthy way. Be in total control of your thoughts, total control of your reaction to life.

Wisdom is knowledge, ability, insight, good sense and judgment for a course of action. Wisdom also means accumulated philosophic or scientific learning.

Understanding flows in all relationships when there is love. Trust in yourself deeply and completely to guide your thoughts and actions to paths of happiness. You should have the capacity to follow through on everything you want to create, including a new way of living. Life is always ready and able to work through you. Love is an essence, an atmosphere of thought. Love defies analysis, as does life itself. Love is spontaneous and impersonal.

Love is a mental as well as a physical state. When you know this in your heart and mind, love then works in perfect peace. Always expect the good, have enthusiasm and, above all, have the state of consciousness at all times of Control Thought.

To the degree that you perfect your thoughts, the perfection in all men will appear to you. For you to have unconditional love that lights the pathways in your life, the most important thing to remember is that you are always causing something to be created for you. What you concentrate on with your thoughts then becomes the effect of the causation put into action.

Thoughts are images that you see in your mind or feel as the result of your thoughts.

Thoughts are never idle; every thought you have brings peace, war, love or fear. Every thought you have contributes to truth or illusion.

Making your mark on the world isn't about dressing well, saying the right things, or taking the perfect action; it the overall impact of your character that you show to the world.

Working and living together in harmony and love for other human beings is the only way that peace can be enjoyed. Tolerance is the key ingredient to enlightenment and wisdom. Realize your power and create your life accordingly. It's important to believe the best can happen to you. Much better than a vacant mind is one filled with spiritual thoughts. The results of your thoughts are the best interpretation of you. Know that you are going to make a difference because you have the power within you.

If you lack love, it may simply be that you have not found it yet. Don't stop looking for it – some one is waiting for you! **The secret is within you, and there is no lack when you discover who you are.**

MEDITATION

M editation is contemplation or reflection while you focus your thoughts on an intended purpose. It is a plan or project in the mind through which you gain mastery. Meditation is quite similar to a good physical program.

You know by now that your thoughts will produce results only in accordance with your beliefs and that you must give yourself reasons you can accept in order to attain your goals. By replacing an undesirable thought pattern with a desirable one, you will form a new habit of being peaceful on the inside as well as externally.

The mind loyally and impartially carries out the orders given by thoughts, words and deeds. So the mind will recreate that pattern established by a single impression, making an idea or experience on an ever-increasing scale until it dominates.

Only when you change your consciousness of thoughts to a more positive pattern can you expect an undesirable pattern to change permanently. With repetition, a new thought pattern will result in the desired change.

Through the practice of meditation, you may access more of yourself and become closer to yourself and to reality.

As with meditation, a good physical program requires hard work. Neither exercise nor meditation has an age limit. Training the mind, the way an athlete trains his body, is one of the primary aims of meditation. By meditating, you can produce results.

The creative power is the thinker behind your thoughts. First you must have the willingness to allow change to take place. Any situation can be changed when you change your attitude and behavior. This presence of life within accompanies you on your path in life and will always be with you. You can not change the past, but you can change the present, which is the NOW. I once was told by a friend, "I want to be all I can be, and I want to have not a job but an adventure." To me this says it all: to be happy no matter what kind of work you do, make it an adventure.

Life is to be lived and in doing so, you have choices to make day after day. You can be miserable, or you can choose to have a happy life. So make your choices an adventure in your everyday life.

You have to live your own life, have your own experience and recognize your higher power. If you have a thought pattern that is not making you happy, then it is up to you to create a new thought pattern that will bring you the life you want. The universal law of the mind translates your desires into reality. The law is the reflection of your own thoughts.

Awareness of this possibility provides you with a powerful tool for change. There is something in the human condition that yearns to reach the highest level, throwing off the pain and loneliness of separation and experiencing unity with others. In meditation you connect with the one power.

The world is not creating your experience; your thoughts are. Your blood supply is carried to all parts of the body, including your brain. The thoughts you send to your brain are then sent to all parts of the body. There is intelligence in every atom, cell and nerve. Realizing this enables you to send thoughts of love to your body. Love is your ability to desire that only good comes into your life. So in choosing to be healthy, know that your body achieves the thoughts' expectations.

Meditation is the breakthrough in consciousness. It's the awareness of a higher power that will guide you through your life. Meditation actually is using life's law and love, the power of the spirit within. When you meditate, you'll have an inner experience of peace. Through the awareness of the mind, which is the creative outlet, each moment is built upon your thoughts. **You set the power of thought into motion, and the power of your thoughts will bring into your experience whatever you desire.**

Your awareness of this process and your ability to use it can transform the way you look at life. This applies in everything – health, personal, relationship, business and financial activities. Rather than struggling with outer circumstances, you can work with your inner causation.

By spending time alone, breathing deeply and quieting your thoughts, you can improve anything in your life. Meditation will help you awaken your inner guidance. You will recognize the power is within. View your circumstances and, if not happy with them, choose to stop thoughts from these circumstances. Instead, create a new series of thoughts. This will start a new series of circumstances for you.

AUTOSUGGESTION AND CREATIVE VISUALIZATION

Let's consider two more good practices, autosuggestion and creative visualization. Autosuggestion is what you say to your inner self, as positive affirmations. Visualization involves imagining, as clearly and realistically as possible, what you want to happen, as if it were already the case. You create the inner experience of what it would be like to have your desires come true. Close your eyes and imagine your desired goal as already achieved in your life. As for affirmation, it is the act of affirming a positive assertion. The best way to do this is by writing out your affirmations and reading them daily.

Here are some tips to help you practice these techniques. In meditation, breathing is very important. It will rejuvenate your life, but you must remember that it will take time to gain results from meditating.

Improper breathing deprives your brain of sufficient oxygen. This can be caused by breathing from the chest alone, instead of deeply from the abdominal muscles. Before attempting to meditate, I would recommend you practice breathing deeply from your abdominal muscles. Hold your breath and count slowly to three, and release your breath to a count of three. Counting your breath in meditation is essentially designed to teach you how to breathe from your abdominal muscles. You need to practice your breathing and your counting before you can master the art of meditation. It also teaches you to train your mind to do just one thing at a time.

Start by placing yourself in a comfortable chair so that you will have few distracting signals from your body. The goal is to think of nothing but counting your breaths. It will take time, but you will master it. If other thoughts come to you, simply bring yourself back to your counting. It takes practice to be aware of counting and nothing else.

Fifteen minutes is sufficient in the beginning for those who are just starting to meditate. You can add more time as you progress.

The most critical element for your meditation program is being realistic about how much time you will need to spend. You must be consistent and persistent. A good meditation is focused on a particular problem. Meditation, then, occurs when the mind and spirit become one and create the power to change your reality.

After many practice sessions, you will be able to meditate on the first thing you want to change in your life. During meditation, your brain wave pattern changes to the tranquil alpha range, an altered state of consciousness which allows you to discharge the tension of the day. After your meditation, you are refreshed and relieved of symptoms caused by tension. After you have disciplined your mind and body to go into deep meditation, the change you want will occur. This is true of anything you want to change. If you want a healthy body, then you would meditate on health and focus all your thoughts on what needs to heal in your body.

I'd like to tell you a story about myself and the power of these four vital practices, meditation, affirmation, autosuggestion, and creative visualization.

At one time I had a very bad case of arthritis. The doctor gave me a prescription, but after months and months of visits, I didn't improve. There were nights when I could not sleep because the pain was so great. At the worst point, I said to myself that if I have to live with this pain, then I don't want to live. But I thought there must be something I could do. So I began to do research on arthritis. I found out in the medical books that there was no real cure for the disease, so I decided to prescribe my own treatment.

I embarked on a seven day fast, drinking only water and fruit juice without sugar. (I believe that sugar is a major contributor to arthritis.) At the same time, I practiced meditation, autosuggestion, visualization and affirmations.

I meditated on healing my body. In visualization, I mentally brought a healing light to the parts of my body that had pain, and then, while holding my hand on that part of the body, I visualized the light as fire. In addition to meditating and visualizing to heal my arthritis, I also wrote a few affirmations that I read every day:

MY BODY AND ALL ITS ORGANS, TISSUES, MUSCLES AND BONES ARE IN THE PROCESS OF HEALING, TRANSFORMING EVERY ATOM OF MY BEING.

Another affirmation is: I AM NOW HEALING MYSELF. I AM ENERGIZED, ALIVE AND FILLED WITH RADIANT HEALTH!

All these practices worked for me. I have been without pain for many years. What I have written in this book about healing yourself will work if you allow the time needed to heal.

AUTOSUGGESTION

Autosuggestion can change what is stored in the subconscious from negative to positive. For example, if you say to yourself, "My memory is good in every department. I shall always remember whatever I need to know. I shall retain information automatically and with ease. Whatever I wish to recall is immediately present in the correct form in my mind. I am improving rapidly every day, and my memory is better than it has ever been."

"Every day I am becoming more and more lovable and understanding. I am now becoming the center of my world. This happy, joyous mood is now my normal, natural state of mind."

"I establish a major premise in my thinking: that the infinite intelligence of my subconscious mind is guiding, directing and prospering me spiritually. In this way my subconscious mind will automatically direct me wisely in my investments and heal my body."

CREATIVE VISUALIZATION

The visualization technique is by far the most reliable tool you can use to see and realize a new and more prosperous future for yourself. In creative visualization, you focus on specific areas of your life, one at a time – relationships, finances, health, home, possessions, self-esteem. You can create your own categories.

The secret of creative visualization is to imagine as clearly and realistically as possible what you want to happen, as if it has already happened. This inner experience allows you to feel what it will be like for your desire to come true. Close your eyes and picture your desired goal. Put yourself into the picture. Working with all four steps, you will ultimately incorporate them as a part of your natural way of thinking and living.

AFFIRMATIONS

You cannot transform outer matter without transforming your inner matter, for its origin is always the same. There is but one nature, one world and one matter, and so long as you go about change the wrong way, you will not arrive at the results you desire.

Realize the strength within you; bring it forward so that everything you do may be not your own doing, but the doing of that truth within you. All that you can become and do and hear in the physical life is prepared behind the veil within you. Therefore, it's of immense importance to be aware of what goes on within these domains, to be master there and to be able to feel, know and deal with secret forces that determine your destiny and your internal and external growth. You can cure nothing outside if you don't cure what's inside; you cannot master the outer world if you fail to master the inner world.

I now take time to make sure that my thoughts are at peace.

I take charge of the money in my life.

I welcome more and more love into my life.

These are a few affirmations. There are many more from your own situation. An affirmation is something you affirm to happen. It's a positive assertion of things you would like to come about, as well as a solemn declaration of what you want in your life.

PATIENCE HAS POWER

The ability to be patient affects everything you do. All doubt and fear must go, and in their place must come faith and confidence in order for you to be led by the spirit into all good. Having patience implies an element of letting go of worry and concern, and trusting in the law of the mind to create the things you desire.

Patience means so much more than putting up with something or someone. Think of patience as the steadfastness needed for growth to occur. When a seed is planted in the ground, a certain amount of time must pass before it sprouts and grows.

Similarly, when you plant an idea in your mind, you must be patient while the realization of your desire takes place. I often think about the great inventors of our times, such as Thomas Edison and Alexander Bell. They must have had tremendous patience through the trial and error portions of their accomplishments. No doubt they experienced times of discouragement and despair. They created new successes out of old failures and never gave up. Patience indeed does have its reward in the ultimate realization of success.

Recognize and practice patience in every area of your life. Let go of striving to make things happen; let go and let the power of Control Thought reveal through you the desires of your heart. Accept new ideas with an open mind. Don't worry about what people might say or think about you. Your time is far too valuable to be annoyed about what people might say regarding your endeavors. Be concerned only with the constructive side of life.

Self-condemnation is a great barrier to progress. Don't condemn yourself. Through Control Thought you can heal yourself from any limiting belief. It is a wonderful experience and a great adventure to make conscious use of the law, to feel that you can plant an idea in the Mind and see it gradually take form. Have faith and know that there is a great creative power that flows through you. Take time everyday to see life as you wish it to be, to make a mental picture of your idea. Then pass this picture over to the law and go about your business with a calm assurance that within your mind something is taking place. You are surrounded by limitless mind-energy, out of which everything is made. The nature of the universe is always to take form according to a pattern, through the process of the law.

Through your belief and conviction, you provide the pattern for the manifestation of the law. You may often be helped by realizing that the universe is infinite, that you can always draw from it for any desired good, and that there is no limit to the good which can be manifested. When you intellectually understand the nature of the unseen part of our universe and the way it works, and wholeheartedly believe in it, then you can use it more effectively.

Everything in the material world is manifesting according to a pattern. It all begins with an idea, and from that idea matter is then created. Everywhere you look, man has created the world around us. There is a power in the universe that is greater than any individual. You can call it whatever you want, but there is overwhelming evidence that this power does exist. The basic teaching of religion is that there is an infinite power in the universe. And this power is accessible to all, to the degree you believe in it.

Since the power in the universe is everywhere, this presence is in and through you. This presence manifests itself in and through all forms, all people, all conditions. Everything responds to you at the level of your awareness of it. Regardless of what your problems may be, you can turn to this presence. Know that you can rely on this power of the universe to guide you and inspire you to a new life.

The divine presence and power is always with you. Re-program your beliefs and become that person you always wanted to be. You have to remember that your programming determines your beliefs. Re-programming is basically Control Thought. It's up to you to program new thoughts about how you want your life to be, and as you change the inner attitude of your mind, your subconscious mind responds.

Thoughts are things which define a state of consciousness as they become subjective, operate through a creative field, and tend to manifest themselves in form. The perfect universe looks only at what you want, so never limit your view of what is possible. Place no limit on principles, principles being life.

Only through a change in consciousness can the outer condition be altered. Obviously, this is a long-range goal. You are not going to achieve this in a day. But you must challenge yourself to press on toward high consciousness, and in doing so, you will master Control Thought. The most important thing is to face in the right direction, moving forward. Don't resist change.

To have patience, you must learn to accept all people as they are, regardless of color, race or creed. Be patient about their different opinions and do not judge. Devote time each day to get into a regular routine of thinking Control Thought.

You have to practice an attitude of patience. For with patience, you will master all Control Thought. Remember, be patient with others and do not judge what they say or do. Even if you feel they are wrong, they may be right in their stage of growth.

The one wholly true thought you can hold about the past is that it is not here. To think about it at all is therefore to think about illusion. The mind is actually blank when it does this, because it is not really thinking about anything. Your mind can only grasp the present – the only time there is. The mind, therefore, cannot understand time, and cannot, in fact, understand anything until a thought passes into it.

Learning to listen to what kind of thoughts you are having will become a new way of living. All thought tends to create its physical correspondence. For example, you can think yourself into being happy or unhappy; it all begins with a thought. With Control Thought you can turn on the power of enthusiasm for whatever your desires may be. Believe in yourself. Believe in your ability to give yourself successfully to whatever is really important to you. Know this is true.

If you are not experiencing the good you desire, you need to change your thinking. You are able to live your life free of fear, hatred, judgment, lack and limited thoughts, for your mind is limitless. With Control Thought everything you do will bring inner peace.

It's been estimated that we have about 280,000 thoughts a day. Now, I am not suggesting that you have to master all of them. The thoughts that you should control are those about family, health, love, happiness and success. I believe that we have only our thoughts to thank for where we are today. **Remember, you are what your thoughts are.** Also, what the mind can conceive and believe, the Mind can achieve.

Each moment is spent in the Now. Spend that moment in total control, and that moment will be filled with happiness. Your thoughts do two things for you, prompting you to speak and to act. The mind takes care of everything else.

Be of good humor, always think of the future and have no fear. Perseverance will bring happiness. Patience and perseverance lead to success (and a sense of humor helps!)

It's been said that truth will overcome evil. This is true with Control Thought – positive thoughts can and will take over all negative thoughts.

Today is a good day to start weeding out unwanted thoughts and planting the Control Thoughts you wish to see flourish. You need to take stock of the thoughts you are entertaining in your mind, and if there are any that do not promote a feeling of good, let go of them and replace them with Control Thoughts of wholeness and happiness. It's never too late to change. With Control Thought you can change, and as you change, so does the world you live in. **The Now is all you have, and the thoughts you have today create all your tomorrows.**

Each of us has been given dominion over our thoughts, words and actions. Therefore it is not necessary for us to continue in a way that limits us. Each day you can choose to change anything that is not right for you.

Growing in wisdom is actually a letting-go process. Let go of all beliefs or fears that hold you back from experiencing the magnificence that life is offering you.

You must turn your thoughts to the moment, for all you have is the Now. You can maintain a positive statement of truth in everything you enter into, each new experience, certain that it will lead you to your highest good. In aligning yourself with an unfailing principle of truth, you keep your thoughts centered.

Your thought has created your world, and with those thoughts you can make the choice to fulfill all your dreams, all your needs, and all your happiness. You can be free from the past by no longer clinging to thoughts of what was or what might have been. **Live joyously now and fill your life with experience that enriches you.**

It's all up to you.

THE WORLD BEFORE YOU

The world before you accompanies you throughout your journey in life. To make life a great deal easier, bring along Control Thought so that you'll be able to master all your thoughts about what happens to you. Control Thought is your consciousness and your awareness of being in the Now. It's learning to listen to what kind of thoughts you are having and controlling them. Are they positive or negative? Recognize your Mind as the powerhouse, the link to all the power in the Universe. Know that all your thoughts are what govern you along your journey from childhood through adolescence to maturity.

THE POWER OF ATTITUDE

Your life is not determined by what happens to you but how you react to what happens; not by what life brings to you, but by the attitude you bring. A positive attitude causes a chain reaction of events and outcomes. It's a catalyst, a spark that creates extraordinary results.

Attitude is a mental position with regard to a fact or state of Mind. The attitude you have can make your life full of wonderful experiences. What is your basic attitude toward your life? This is an important question for you to answer since your attitude affects your life so profoundly and also because your attitude will be with you for your lifetime. Happy people tend to have a healthy attitude of self-acceptance, a feeling that they deserve to be peaceful and happy. Those who are not happy tend to display the attitude that they are not deserving of life's goods. An attitude that is destructive to your happiness is the belief that you do not compare favorably with others. There is a way to offset this feeling of inadequacy, and that is by radiating love and goodwill to everyone around you. All life moves through the present, so action in the present is the most important. Remember that attitude follows you throughout your life. It's like learning to rule in your life from the time that you can think on your own. It shows the world just what you are all about.

BEHAVIOR

Some people continue to change jobs, mates and friends, but never think of changing themselves. Behavior is what a person does, thinks, feels and believes all through life. Become enthusiastic about new creative ideas and continue to learn and grow. In this manner you will remain young at heart, and your body will reflect your thinking at all times. Behavior is seen by the people around you; you can change your behavior any time you change your thoughts to a more positive direction. Each of your experiences is indelibly written on your consciousness. You can't wipe the record clean, and in most cases, you wouldn't want to. The trick is in finding an empowering way to view the experience. Behavior is action, and action causes things to happen. With Control Thought you can control what your behavior will be at any given time.

EDUCATION

The Mind is like the stomach. It is not how much you put into it that counts, but how much is digested. A hundred mistakes furnish you with a liberal education, if you learn something from each one. The ladder of life is education, which leads to understanding. The second part of life's ladder is concentrating on making a living. The third part of life is giving all our attention to living life, realizing that we are on an endless ocean of life. Retirement is still another step forward on the ladder of life toward wisdom.

COURAGE

More twins are being born these days than ever before. Maybe kids lack the courage to come into the world alone. Courage is being the only one who knows you are afraid.

Courage is as much mental as a moral strength. It's the strength to resist opposition, danger or hardship. Courage enables you to hold your own and keep up your morale when opposed or threatened. Most important, courage is the unwillingness to admit defeat.

To live your life the way you would like, you must have the courage not to worry what other people think of you. Their thoughts about you are not your concern. Your life belongs to you, so live just the way you feel, knowing that you are doing your very best.

As a rule, we view life with amused intellectual detachment, but we can be roused to a towering fury on two subjects: courage and cowardice. Without courage one can never achieve big goals. Difficulties happen to everyone, but to recognize what is happening, take charge and see it through requires courage within you. To the average person this courage can be of great value.

REVENGE

Getting even does not solve anything. Revenge is a response to feeling hurt, and it is a very primitive response in search of justice. But in fact, the person who concentrates on revenge is not able to cope with the hurt.

On the other side of the pain there is relief and great peace of Mind when you forgive everything that has caused you pain. The thoughts you have while thinking of revenge are all negative thoughts. You can change your thoughts into forgiving, positive thoughts. You'll find as each day goes by that the hurt slowly leaves your body, and then you are free of it.

VISION

From the first chip off the marble or the first stroke of the brush on a canvas, the artist pictured an inner vision. When you look at a great masterpiece, you know that the artist must have held to an inner vision of beauty and form while completing it. **You too can be an artist who creates masterpieces in your life.** Your inner vision of health can affect your body in positive ways. Your thoughts reveal the life that is within you. Your inner vision of peace shows forth as harmony in your relationships.

By holding to your inner vision, you are allowing the peace within to come forward in all that you are and all that you do.

GUIDANCE

You may ask yourself, "Am I experiencing the kind of problem that I don't seem to have any solution for? Do I feel defeated even before I begin searching for an answer because the situation calls for more understanding than I think I have?" The answer to these questions and others comes in tuning in to that power which is within. Using Control Thought aligned with the Divine Intelligence and the Divine Understanding, which are always available to you whenever you need them, the answer to any situation will be clear.

THE CHANGE YOU NEED

The change you need to make occurs first within yourself. If you are not experiencing the good you desire, you need to change your thinking. For instance, if you don't like what you are hearing in your mind, or not getting what you are looking for, now is the time to change all the thoughts that keep you from getting what you want in life. With Control Thoughts you can change the way you are living and achieve whatever you would like to experience. As I keep reminding you, your life will reflect the thoughts you choose.

FEAR

Every attitude contains the seed of a corresponding thought or experience. We have many fears in life. To overcome fear, whatever it may be, say to yourself slowly and quietly and positively, "I am mastering this fear, I am overcoming it now. I am relaxed and at ease, and I know that it is so." As these positive seeds of thoughts sink into the subconscious, they grow and you become poised, serene and calm.

Suppose you have a sudden fearful thought about catching a cold or about the security of your job. Deal with the thought and dispose of it immediately. Don't procrastinate and don't give in to it. Most fears are actually generated by too much reading, thinking and talking. A young woman or a mother who reads the extensive literature about the young might became fearful about how to raise her children.

Overcoming fear is a very elementary process. **We overcome fear with action.** Fear is nature's warning signal to get busy getting your life in order, by changing your thoughts to a more positive way of thinking. Remember that fear can be overcome – you just have to work at it. Change fear into faith. People desperately need faith. In these urgent times, it sometimes seems that civilization itself will fail. This is a spiritual problem because people around the world have lost the ability to love those who are not like them. To counteract what we see in the world, we need to have faith. We shall win because we have the preponderance and resources of faith.

HAPPINESS

You have to know intuitively that lasting happiness is not dependent on outer sources or conditions. It is only through the recognition, development, and the use of your inherent qualities that you can find true happiness. Only through developing positive thoughts about yourself, the world around you, and the people in your life can you have happiness in your life. **This is the beginning of happiness – it is found within.** All the thoughts that you have about being happy will produce happiness in your outer life.

It's like anything else in your life that you would like to master. Control Thoughts will take you there. You have to remember that it does not change overnight; you have to work at it constantly. Changing your thinking toward a more positive direction is the first step you must take.

EGO

Your own state of mind is a good example of how the ego is formed. For ego is part of everyone's mind. When you feel at a loss, with things not going the way you would like, then the ego in your mind comes through you. The ego tries to control what it doesn't like and to change circumstances to the way it wants things to be. Your ego can run your life if you don't take control of it; it can cause you a lot of trouble. Your ego arises from a sense of separation, and if you continue in that belief, you remain separated. Your own power along with your Control Thought can guide you to control your ego.

HOPE

I believe that the possibility for desire and hope is within everyone. Acceptance is the first thing we need in any situation in which we hope to find the answers. Hope is the intention of the good that you are asking for, hope for health, friendship, family and wise choices.

Hope is a desire accompanied by expectation, a belief that someone or something will meet with success. Suppose you ask a person in the hospital, "How are you today?" and he answers, "I'm all right, don't worry about me." This is the hope of a sick man. Through sheer will power, hope, and faith, he'll be OK. Many people believe that the spirit within can make their hope come true.

When situations seem incapable of a successful outcome, you must not let hopeless thoughts enter your mind. Believe in your ability to hope. Establish a purpose and increase your effectiveness through your Control Thought. Hope is an intention and a desire. Cherish the desire with expectation of fulfillment. With desire, trust, and hope these dreams can come true.

ACTION

You can't get anywhere unless you start. Don't forget that people will judge you by your actions, not your intentions. You may have a heart of gold, but so does a hard-boiled egg. Begin where you are, but don't stay there for too long. The best time to do something worthwhile is between yesterday and tomorrow. What you say and do shows what you are. Failure always overtakes those who have the power to do things but lack the will to act. The only thing you have to fear is not doing something about your fear.

LOVE IS A MENTAL AS WELL AS A PHYSICAL STATE

Know that to the degree that you perfect your thoughts, the perfection of all men will appear to you. For unconditional love lights the pathway of life. One of the most important things to remember is that you are always causing something to be created for you. The energy of thought through the mind is always producing the universal law. The law follows the thought, thought follows desire. In the physical world a strong affection for another inspires feelings of love and desire.

YOU CAN BE HEALED

When you are conscious of perfect health, the body is whole. To be healthier than you are now, start doing your part to consistently reverse all negative thoughts of sickness and disease. You can be healed when you make contact with that inner being and have faith that your healing will take place. Thoughts are the building blocks you are using to create your health. Limited thoughts oppose the realization of health. The negative thoughts need to be controlled.

Meditation on healing and health is very important, as are auto-suggestion, visualization and affirmations. All of these help to heal whatever you want to change. Thought energy is the force generated in the brain by the process of thinking. Your thoughts travel from cells in the brain to their destination in that part of the body that needs to be healed.

MEDITATION IS SIMILAR TO A PHYSICAL PROGRAM

Both require hard work and, like exercise, meditation has no age limits. Training the mind the way athletes train their bodies is one of the primary aims of meditation. During meditation you focus your thoughts on an intended purpose, knowing that your thoughts will produce results in accordance with your beliefs. You must give yourself reasons you can believe in regarding your ability to attain your goals.

FORGIVENESS

Why is it sometimes difficult to forgive others? Is it because what they did was so unforgettable? Or because you have become attached to the negative memories by continually thinking about them?

Releasing the past is the first step toward complete forgiveness. You can let go when you realize that the words and actions of others are a response from their own beliefs – their responsibility, not yours. You are growing and unfolding at your own rate and no two people think or feel exactly the same way. Knowing this, you release the need to have everything happen the way you think it should. There is nothing you or anyone else can do that cannot be forgiven, regardless of the past. Every day can be a new beginning.

LOVE

Stop searching for love outside yourself. Love as you would be loved. Become aware of any negative thoughts that are not in tune with the infinite beauty of nature. Everything works in the universe by law, and when love is all, two become one. You are only at the beginning of an eternal journey, and all along the way you should respond to the highest healing power of love. When you have love in your heart for all, you are motivated by love from within and empowered by it. Everyone deserves to have love. Love keeps open the channels of communication in all relationships and then love flows into your life.

ARGUMENTS

For some unknown reason, people disagree and argue in the attempt to persuade or convince the other. Quarreling causes discomfort and often ends with unpleasant, disagreeable feelings. In an argument the best approach is to walk away, just as our mothers used to advise us. Then after a time think about what caused the argument. In any argument things are said that can hurt; once said, they can never be taken back.

THE GREATEST PHYSICIAN IN THE WORLD

The power of the mind is what causes your heart to beat, your muscles to have strength and your body to heal. Feel good about yourself and you will embody the image of perfect health. Your everyday thinking about your health must be positive at all times. There is a state of consciousness which can heal, and this state of consciousness can be obtained if you want it and are willing to work for it. It is possible to have a true subjective concept of health; you have within you the greatest physician in the world with the power of the mind.

DOUBT

To become a living embodiment of success, your belief in yourself must be so firm that doubt can not short-circuit it. You have to let go of thoughts of doubt, distrust, worry, condemnation and fear, replacing them with positive thoughts. When you say "I will try to do that," by saying "I'll try," you have already accepted defeat. What you are really saying is either "I don't know if I can" or "I don't know if I want to." Both of these statements denote doubt, and doubt is a giant obstacle to the manifestation of good. Instead, say "I will do that" – a positive statement.

YOU MUST TAKE ACTION

You must take immediate action – stand up and speak words of peace amidst the storm of human thoughts. Remember that all you can ever do is either support or attempt to suppress one another. You can engage only in unity or division. Inevitably, whatever you promote, you produce.

THE WONDERFUL THING ABOUT THE MIND

is that you can change your input to create a different result. Sounds simple, and it is. If your belief systems have been held in place for many years, now is the time to reprogram a whole new input and be faithful about practicing your new thought pattern.

LIFE

Life is a continuous process of growth as you move from one lesson to another. No one stays the same as we were a few years ago. Change happens to all of us, day after day. Yes, life is a now experience, and there is a law that responds to whatever you are thinking. You can change your thoughts, one by one, which means you can stop playing at life and start living life the way you would like.

Disappointments do appear now and then. Disappointment can be a sign that we have yet to fulfill a dream. This is the reality of life. What needs to be done when this happens is to get back into thoughts on how to be successful.

By using Control Thought, you'll be able to limit disappointment and grow from it.

YOU ARE UNIQUE

Yes, you are living a life that is unique because there is no one else in the world just like you. You contain within you all the intelligence and power that you need to live a happy life. Know that you have the ability and the power at your command to express life in peace, happiness, abundance and satisfaction. This power within you will act on whatever you say you can do. Each day you are growing in wisdom and understanding. You are a part of the Infinite life.

Control Thoughts can bring wonderful and rewarding experiences to your life. So never doubt your individuality, and hold your head high.

INNER PEACE

Live in the present and enjoy the fullness in every moment. Have that moment be filled with positive thoughts. Learn to be guided by intuition rather than externally imposed interpretations of what is good or bad for you. With the power of thought, you can have the power to heal anything, and you can turn stress into strength. There is an affirmation I say to myself when things are not going the way I would like:

I NOW TAKE TIME TO MAKE SURE THAT MY THOUGHTS ARE AT PEACE. I ESTABLISH A MAJOR PREMISE IN MY THINKING: THAT THE INFINITE INTELLIGENCE OF MY MIND IS GUIDING, DIRECTING AND PROSPERING ME SPIRITUALLY AND MENTALLY.

We actively decide for ourselves to have inner peace.

CAUSE AND EFFECT

The law of cause and effect is very simple and inevitable. Cause and effect are inseparable. You should think of them that way.

Thought is the cause, action is the effect. The law of cause and effect transcends a higher use of the law. The law of cause and effect operates on your beliefs as they actually exist. You have been endowed with a creative mind, whether or not you know it.

Emotion and reason should be in balance in your life at all times. Like cause and effect, balance is possible within your thoughts in all situations. Cause and effect will be with you your whole life. So learn to control your thoughts with the Power of Control Thought.

CHOOSE TO THINK POSITIVE THOUGHTS ABOUT YOURSELF AND YOUR ENVIRONMENT

What you experience today is the direct result of what you have chosen either consciously or unconsciously in the past. You can therefore create what you want to experience now. With the use of Control Thought, it becomes easer to change things that are not what you want. Being in control of everything you go through is a great thing.

Choose positive thoughts as you go through life, and you will experience the difference.

THE MORE YOU AFFIRM THE IDEA OF HEALTH, THE MORE YOU GRADUALLY MAKE A CHANNEL FOR THE HEALING POWER.

There is no condition that cannot be overcome. Hold to this conviction. Believe it, affirm it, and know that the more you hold to the spiritual idea of health, the stronger you become in faith. The more you affirm the idea of health, the more you gradually make a channel for the healing power.

Just what is health? Health is a condition of being sound in mind and body. Health is freedom from physical disease or pain. Health is not only of the body, but also of the mind and spirit.

The wonderful thing about the mind is that you can change your input to create a different result. Quietly contemplate perfect health within you right now. This health flows IN HARMONY AND IN BALANCE THROUGHOUT YOUR MIND AND BODY. Your thoughts are the creative avenues within you, and since thoughts are creative, they can heal disease.

RICHES

Think and believe you can move from the impossible to I'M POSSIBLE.

People have thoughts of something that they want – for example a new car – then in the next breath they say, "But I can't afford it." There again is "I can't." When it comes to weight loss, people deny that such a thing is possible for them because of various negative reasons. What is the ultimate result? Nothing happens. Yes, life is a now experience and there is a law that responds to whatever you are thinking. The more spiritual the thought, the higher its manifestation. Spiritual thoughts require an absolute belief and reliance upon the truth.

Maintaining a burning desire transforms those thoughts into concrete action. Wishing will not bring riches, but desiring riches with an obsession and planning definite ways and means to acquire riches takes action. **The world is filled with an abundance of opportunity.** Limitation and poverty are the results of restricted ways of thinking.

RETIREMENT IS A NEW VENTURE

It is a new challenge, a new path and the beginning of the fulfillment of a long dream. Look at it this way – retirement is a promotion from kindergarten to the first grade. The ladder of life is education toward greater understanding of life.

Retirement is still another step forward on the ladder of life experience, and that step is the gaining of wisdom. Retirement is not the end but a beginning to live freely, mentally and physically. It gives you time to pursue new dreams.

A WORLD OF WONDERS

Before you lies everything in the world – a world that we see, hear, smell and feel, nature with its wonder and beauty, and the great minds of the past and present whose thoughts have created the world we enjoy. Master painters, sculptors, great composers, authors and poets created work all for the world to enjoy. Everything in the Universe is constantly being created, just like you and me. Everything is changing every day with the creation of new and better ways to live. Everything in the Universe has its own vibrations, or energy. This energy is always in motion, being born, unfolding, transforming and shifting.

The energy that we create through our thoughts can be felt by someone near or far away. Beauty can raise your energy when you cultivate an appreciation for it and become aware of the unique and beautiful qualities of nature. When you do that, you raise your vibrations of consciousness. You rise to a higher vibration of energy with awareness, gratitude, beauty, joy and trust. Then you are more in harmony with the Universal energy.

Sometimes your life moves in mysterious ways. When that happens, you need to keep your energy high and remain centered. Life is always moving and unfolding; today you are here, and everything in your experience has brought you to this very moment.

I have asked many times to understand life and its meaning. What I have come to believe is that life is an ongoing process with many ups and downs. Without knowing the reasons, we must have faith and find our own way. The study of Control Thought has helped me to understand life. Like happiness, understanding is found within.

At the doorway for change, all you need to do is walk through for the change to happen. Your opinions about life can change, but sometimes it takes longer to change the opinions you have about yourself. Most people are willing to change not when they see the light, but when they feel the heat. Think right thoughts and act right because what you think and do makes you what you are. Everyone has his own character which he exhibits. Remember, people judge your character by observing what you stand for. EVEN IF THIS IS THE DAWN OF A BRIGHT NEW WORLD, MOST ARE STILL IN THE DARK.

Happiness does not come from what you have but from what you are.

ENLIGHTENMENT

Enlightenment is seeing your world as it really is and watching your way along the journey of life. If you're humble and see the beauty that surrounds you, you will enjoy the ever-changing self and experience a feeling of oneness with others. Time passes rapidly, but you can be happy with thoughts you control. Remember, it's a new beginning, and it takes time to grow. You have to be patient with yourself during all the stages you have to go through.

Take a look at yourself as you are today. This will take you to that inner place where you are spirit, with enlightenment unfolding within you. As you think about that, know you'll need to make some changes in your life. When you accept the changes and realize them, you'll be able to live as an enlightened person. This new way of thinking is The Power of Control Thought. Enlightened, you can be a friend to the world.

You appreciate the beauty around you at all times, seeing your body changing with each passing day. Keep faith that you'll be able to live with very few problems, and enjoy each day because you know that our time here is short.

You can live every day in a state of bliss with the ones you love and and who care for you. I give thanks for the friends who did so very much for me in many ways. And along the way, I helped the ones in need. I was able to create a most positive world around me and the people in my world. The greatest gift we have is our mind to create.

Attitude is all-important. I have said this before, but you need to hear it again. Your attitude accompanies you throughout your life. Everyone you meet along your journey sees you as you are by your attitude. So remember that the mind is the creative principle of life. Don't be afraid to use your creative ability to form your outer world. Don't let the past eat away at you, and don't worry about the future – make the best of the Now. Life is too precious to let it slip by while you're feeling sorry for yourself.

TAKE ACTION AND PUT YOUR LIFE IN ORDER.
ENJOY THE VERY BEST THAT YOU CAN CREATE FOR
YOURSELF.

YOU CAN CREATE YOUR LIFE NOW

Yes, you are in the Now, not in the past or the future. The past is gone, and you need not suffer from yesterday's mistakes or losses. The future is not here yet, and you know not what it will bring. Of course you have to make plans for the future, but all of us have only the Now.

In the Now you can cultivate a number of things:

Things that bring you happiness, like the beauty of nature, playing with your children, or entertaining yourself. Now is the time to discard negative thoughts and replace them with all positive Control Thoughts.

How is one to achieve this, when there is so much confusion, negativity and armed conflict going on in the world around us? For centuries the great teachers, ministers and mystics have told us to turn within to find peace. How do we do that? Turn within means to turn from the outer emotions and confusion to the inner emotions.

This is not always easy. It requires constant alertness and awareness of your thoughts as well as your actions. Take a moment to turn within and quiet the self. Realize the truth is peace and it's always with you.

You are living in the Now age. Now is the time to be present and clear about who you are. Know that you can experience all in the Now. Know that you have created this very moment in time. Take this time and experience your own energy and your inner being. You can't plan your whole life; looking into the future is not being in the Now. Create your happiness today, for happiness is a thought. You may call it Control Thought, an attitude that you show to the world. Cause and effect will always take place. Remember that within the Now there is always the next moment waiting to become the Now. It all comes down to being in the Now.

LIVE YOUR DREAM.

BE YOUR DREAM.

KNOW THAT ALL IS POSSIBLE.

BELIEVE IT, HAVE FAITH IN IT AND
ENJOY LIVING IN THE NOW.

IT'S ALL UP TO YOU.

HAVE A GREAT LIFE AND ENJOY.

BOOKS OF INTEREST

The Power of your Subconscious Mind
By Dr. Joseph Murphy

Wisdom of the Ages and *the Power of Intention*
By Dr. Wayne T. Dyer

Living in the Light and *Creative Vizualization*
By Shakti Gawain

The Seven Spiritual Laws of Success
By Deepak Chopra

ABOUT THE AUTHOR

Roy F. Messier, the tenth of eleven children, was born in Worcester, Massachusetts. After completing school, he served in the Navy for four years and then moved to New York City, where he worked as a hairstylist for twenty years.

After moving to California and studying Culinary Art, Mr. Messier worked in the restaurant business. Next he studied medical and metaphysical sciences at the Universal Truth Center of California. After completing Science of Mind courses in 1990, he served as a practitioner at the Religious Science Church of the Desert in Palm Desert, California, under Dr. Tom Costa. He received his ministerial credential in 1997.

Other accomplishments include reaching the gold level of Toastmasters International, certification by Neuro-Linguistic Programming, and receiving the Editor's Choice Award for Outstanding Achievement in Poetry in 1998.

Currently residing in California, Mr. Messier paints and writes.

NOTES

NOTES

AFFIRMATIONS

AFFIRMATIONS